Table of Contents

Acknowledgments

Implementing Asset Management: A Practical Guide was produced and published by the Association of Metropolitan Water Agencies (AMWA), the National Association of Clean Water Agencies (NACWA), and the Water Environment Federation® (WEF®). *Implementing Asset Management: A Practical Guide* provides a logical, step-wise methodology for improving the assessment of water and wastewater infrastructure, and for making fact-based decisions on capital investments.

Special thanks are extended to the members of the project workgroup: Michael Coffey, Tucson Water, Tucson, Ariz., and N. Max Hicks, P.E., Augusta Utilities Department, Augusta, Ga., for AMWA; Ed McCormick, East Bay Municipal Utility District, Oakland, Calif., and Jon Schellpfeffer, Madison Metropolitan Sewerage District, Madison, Wis., for NACWA; and James E. Patterson, MPA, Columbus Water Works, Columbus, Ga., and Amy J. Siebert, P.E., Department of Public Works Sewer Division, Greenwich, Conn., for WEF®. The publication is a direct result of their dedicated efforts, detailed comments, advice and encouragement.

Review of the document was also provided by Linda L. Blankenship, P.E., BCEE, EMA, Inc., Vienna, Va.; Todd Cleaver, Gwinnett County Department of Water Resources, Lawrenceville, Ga.; Robert G. Decker, P.E., Village of Ruidoso, N.M.; Bruce W. Husselbee, P.E., Hampton Roads Sanitation District, Virginia Beach, Va.; David M. Mason, Infrastructure Management Group, Inc., Bethesda, Md.; Franklin D. Munsey, P.E., Munsey Engineering, Lady Lake, Fla.; Patrick A. Obenauf, P.E., Milwaukee Metropolitan Sewerage District, Milwaukee, Wis.; Steven M. Ravel, P.E., BCEE, Richard P. Arber Associates, Lakewood, Colo.; Joe C. Stowe, Jr., CH2M Hill, Charlotte, N.C.; Michael W. Sweeney Ph.D., P.E., EMA, Inc., Louisville, Ky.; Philip Tiewater, P.E., ELM Consulting, Wayne, Pa.; David C. Vago, J.D., P.E., Wade Trim Operations Services, Inc., Livonia, Mich.; and Randy Weaver, City of San Diego, Metropolitan Wastewater Department, San Diego, Calif.

This document was prepared by CH2M HILL, Inc. under the direction of Alan B. Ispass, P.E., BCEE, Vice President, Utility Management Solutions, under the general direction of AMWA, NACWA, and WEF®.

April 2007

Executive Summary

In early 2002, NACWA (in cooperation with AMWA, AWWA and WEF), released the first comprehensive publication focused on helping U.S. water and wastewater utilities improve the management of their infrastructure assets, *Managing Public Infrastructure Assets to Minimize Cost and Maximize Performance*, commonly known as the "Asset Management Handbook" (*Handbook*). Today, the concern for aging infrastructure persists. Utility managers continue to look for the most effective decision-making tools to assure continued high-quality services and make prudent and rational decisions for investment in infrastructure. This publication, *Implementing Asset Management: A Practical Guide* (*Practical Guide*), directs the theory and elements of asset management detailed in the *Handbook* into a logical, step-wise method for improving the assessment of water and wastewater infrastructure and for making fact-based decisions on capital investments.

The concepts presented in the chapters that follow are applicable to all utilities, whether large or small. The methodology for implementing those concepts is scalable to match the size and resources of each utility. For example, the methodology allows the utility to drill as far down into its infrastructure assets as resources permit. The processes involved can be carried out by just a few staff members or by a dozen; they can use pencil and paper, or software applications. Since the foundation of the methodology is based upon risk – that is, the infrastructure assets posing the highest risk to the utility are addressed first – valuable information and guidance for improved decision making will be realized without evaluating every asset in the utility system.

This publication includes six chapters and three appendices. Chapter 1 provides a synopsis of the impetus for this publication. Chapter 2 updates the state of the country's water and wastewater infrastructure, describing the results of the studies and assessments performed since the publication of the *Handbook* in 2002. Chapters 3, 4 and 5 form the essence of this *Practical Guide*:

- Chapter 3 expands on the definition of *asset management* originally presented in the *Handbook* and explains the 12 key concepts of effective asset management (Exhibit ES.1) and their link to the elements originally characterized.

EXHIBIT ES.1: Key Concepts of Effective Asset Management

Knowledge of:	Levels of service Assets and their characteristics Physical condition of assets Performance of assets Total cost of asset ownership
Ability to:	Optimize O&M activities Assess asset risk Identify and evaluate risk mitigation options Prioritize options and fund within available budget Predict future demands Effectively manage information and employ decision support tools Obtain and sustain organizational coordination and commitment

· Chapter 4 provides an overview of risk and how the risk equation (Exhibit ES.2) is employed to implement asset management in a logical and step-wise manner.

EXHIBIT ES.2 Mathematical Expression for Risk

Risk equation: Risk = [(Consequence) x (Likelihood)]

Examples of how matrices can be used to quantify both consequence and likelihood of asset failure in order to score the relative risk of infrastructure assets are presented. Options for mitigating asset risk are discussed along with a method for prioritizing the risk mitigation options. The chapter closes with an overview of how risk can also be used to plan maintenance programs and how redundancy affects risk.

· Chapter 5 details the step-wise process for implementing the risk-based Asset Management methodology. Examples are presented for quantifying consequence, likelihood and risk, along with selecting and prioritizing risk mitigation options.

The final chapter, Chapter 6, provides a synopsis of the software applications available to assist utilities in better managing their infrastructure and the principles of integrating information systems to most effectively manage the enormous volumes of asset data that can be generated through asset management processes.

The publication concludes with three appendices:

· A collection of sample forms and templates that can be used to implement the step-wise methodology.

· A list of references, both in hardcopy and online, covering a broad range of asset management topics.

· A glossary to guide the reader in the terms used in this publication.

CHAPTER 1

Introduction

Over the last several years, asset management has become a popular topic among water and wastewater professionals in the United States. Faced with the challenges of aging infrastructure, the lack of federal funding, and the desire to maintain affordable rates while meeting customer expectations, utility managers are looking for more effective ways to make decisions about capital improvements and infrastructure maintenance.

In early 2002, NACWA (in cooperation with AMWA, AWWA and WEF), published *Managing Public Infrastructure Assets to Minimize Cost and Maximize Performance*. This "Asset Management Handbook," (*Handbook*) as this publication has become known, was the first comprehensive document focused on helping U.S. water and wastewater utilities improve the management of their infrastructure assets. The Handbook provides an overview of how agencies in other countries are improving their stewardship of public infrastructure assets. It also describes the demographics that are driving the water and wastewater industries in the U.S. to focus on how they will ensure a reliable infrastructure for the future. Most importantly, the *Handbook* presents the fundamentals of asset management and describes the strategic approach to implementation in water and wastewater utilities.

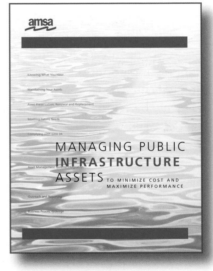

Since the release of the *Handbook*, research on water and wastewater infrastructure assets has increased in the U.S. and other countries, and asset management conferences and training seminars have proliferated. There is nearly universal agreement in the U.S. water and wastewater sector that resources are needed to address the maintenance and renewal of soon-to-be or already deteriorating infrastructure. However, most utilities have yet to implement structured asset management practices or even develop a plan to do so. To date, many utilities have limited their approach to better management of assets to the acquisition of software applications such as computerized maintenance management systems (CMMS), work order management systems, and geographic information systems (GIS). These applications are important tools for utilities to efficiently operate and maintain their assets. However, the benefits can be greatly enhanced if they are incorporated into a system-wide, integrated approach to asset management that includes considerations for risk, levels of service, life-cycle costing, and other important concepts addressed in the following chapters.

Much of the reluctance of utility managers to implement asset management appears to stem from a feeling of being overwhelmed by the scope of such an effort, the lack of needed resources, and the question of where to begin. In reality, asset management does not need to be overwhelming, nor does it require a substantial increase in resources, and there is a very logical place to begin.

This publication, *Implementing Asset Management: A Practical Guide* (*Practical Guide*), provides utility professionals a step-by-step method for beginning a process of continued improvement in the management of their infrastructure assets. The concepts and processes presented are applicable to utilities of all sizes and can also be applied to infrastructure assets other than those in water and wastewater systems. Depending on the availability of resources, utilities can address their infrastructure assets at a broad, system-wide level by grouping assets or drill-down to individual asset components and elements. At any and all points in the process, value is gained through a more structured evaluation and decision-making process.

It is important to keep in mind that effective asset management is not just about infrastructure and technology; it also requires the education and continued professional development of human resources, the integration and verification of information and data, prudent management of finances, and adoption of business processes that are aligned with the organization's mission, vision, and goals.

In reality, asset management does not need to be overwhelming, nor does it require a substantial increase in resources, and there is a very logical place to begin.

CHAPTER 2

Overview of Water and Wastewater Infrastructure Management

Although "asset management" is a fairly recent catchphrase for the water and wastewater sector, the management of assets is nothing new. Over decades, water and wastewater utilities have been managing their infrastructure assets through maintenance activities and capital renewal in order to continue to deliver essential services to their customers. However, much of this management has been informal and ad hoc. Decisions concerning infrastructure renewal have typically been based on the perception and intuition of utility staff members, pressure from stakeholders, and political persuasion. Information and operational data concerning assets is frequently unreliable and possibly unavailable. Different files and databases can contain conflicting information. In many cases the software applications purchased by utilities—computerized maintenance management systems, geographic information systems, customer information systems, and financial information systems—are not integrated, leading to ambiguities and inefficiency. So, while the management of assets is nothing new for utilities, the risk-based and fact-based methodologies of today's asset management strategies are new.

> *...while the management of assets is nothing new for utilities, the risk-based and fact-based methodologies of today's asset management strategies are new.*

Past Approaches to Managing Infrastructure Assets

During the last half of the 20th Century, the focus on adequately maintaining water and wastewater infrastructure was somewhat enhanced as a result of federal legislation such as the Water Pollution Control Act Amendments of 1972 (now known as the Clean Water Act) and the Safe Drinking Water Act of 1974, as well as the federal and state regulations that followed. If utilities had not done so before, these laws and regulations provided significant incentives, both positive (federal and state grants) and negative (enforcement action), to ensure that infrastructure was properly performing in order to maintain statutory and regulatory compliance.

Probably the first widely accepted uniform approach for improved planning of infrastructure assets was the "201 Facility Plan" prepared by wastewater agencies across the U.S. The purpose of this plan was to program the infrastructure improvements necessary to meet the requirements of the Clean Water Act. For many utilities, these "201 Facility Plans," and to a similar extent the "208 Areawide Waste Treatment Management Plans," were the first formal process whereby infrastructure assets were evaluated and detailed plans for meeting regulatory requirements and future customer demands were developed.

Over the past 30 years, water and wastewater utilities have become more aware of the benefits of planning for their infrastructure needs and have at least adopted five-year capital improvement programs. Others have undertaken more long-term planning and have prepared 20-, 25-, or 30-year master plans that provide utility managers with a foundation for projecting financial and operational impacts. However, while utility managers may have become more focused on planning, it is apparent from the general state of the water and wastewater infrastructure in the U.S. that plans are insufficient, not being implemented, or both.

The Current State of the Nation's Water and Wastewater Infrastructure

Evidence from several sources published since the release of the *Handbook* in 2002 suggests, as summarized below, that the integrity of the nation's drinking water and wastewater infrastructure is at risk unless there is a concerted effort to improve the management of key infrastructure assets—treatment plants, pipelines, and other facilities—and a significant investment is made in maintaining, rehabilitating, and replacing these assets. Adding to the problem is that many utilities have not been generating enough revenues from user charges and other local sources to cover the full cost of service.

GAO Reports

The U.S. Government Accountability Office (GAO) reported in August 2002[1] that more than one-third of utilities had 20 percent or more of their pipelines nearing the end of their useful life; for 1 in 10 utilities, 50 percent or more of their pipelines were nearing the end of their useful life. The same report indicated that 29 percent of drinking water utilities and 41 percent of wastewater utilities were not generating enough revenue from user rates and other local sources to cover their full cost of service. Additionally, GAO reported that approximately one-third of the utilities deferred maintenance because of insufficient funding and lacked basic plans for managing their capital assets.

...the integrity of the nation's drinking water and wastewater infrastructure is at risk...

EPA Needs Surveys

The EPA's Office of Wastewater Management conducts the Clean Watersheds Needs Survey (CWNS) on a periodic basis to provide Congress with an estimate of clean water needs for the U.S. As reported in the CWNS 2000 Report to Congress, completed in 2003, the current financial estimate to meet the clean water needs for the nation is $181.2 billion, a $26.6 billion increase from the previous survey completed in 1996. Likewise, the EPA's Office of Water regularly conducts an infrastructure needs survey and assessment and also reports its findings to Congress. EPA most recently conducted a Drinking Water Infrastructure Needs Survey and Assessment in 2003 and found that the nation's water systems need to invest $276.8 billion over the next 20 years to continue to provide safe drinking water to their consumers.

ASCE Report Card

The American Society of Civil Engineers' (ASCE) *2005 Report Card for America's Infrastructure* updated its report card that was first published in 2001. The score card indicates that the U.S. water and

---

[1] GAO-02-764 *Water Utility Financing and Planning*, August 2002.

wastewater infrastructure has shown little to no improvement since receiving a collective D+ in 2001, with some areas sliding toward failing grades. Wastewater infrastructure received a D- grade. In its report, ASCE cited the EPA's gap analysis that assessed the difference between current spending for wastewater infrastructure and total funding needs. Drinking water scored no better, also receiving a D- grade. ASCE stated that the nation faces an annual shortfall of $11 billion to replace aging facilities and comply with safe drinking water regulations.

Unfortunately, the aging infrastructure and the general lack of funding available to adequately maintain and renew pipelines and facilities are not the only challenges that water and wastewater utilities face today. Utility managers must also address the security of their systems, the changing demographics of their workforce, new regulatory requirements, and the unpredictable economy with currently increasing interest rates.

Benefits of Asset Management

While not a panacea, employing good asset management practices can go a long way in helping utilities address their numerous challenges. Because water and wastewater utilities are among the most capital intensive industries, requiring up to three times more capital to generate a dollar of revenue than electric utilities[2], focusing on improving the management of capital infrastructure assets is also a wise investment of resources. Numerous benefits have been realized by utilities that have already begun to implement asset management practices, as related in the 2004 GAO report titled, *Water Infrastructure - Comprehensive Asset Management Has Potential to Help Utilities Better Identify Needs and Plan Future Investments*[3], the *International Infrastructure Management Manual*[4], and in articles published in the industry media.

Because water and wastewater utilities are among the most capital intensive industries, requiring up to three times more capital to generate a dollar of revenue than electric utilities, focusing on improving the management of capital infrastructure assets is also a wise investment of resources.

Several benefits that may be achieved through the adoption of asset management concepts are discussed below.

Rigorous and Defensible Decision Making

The traditional decision-making approach for developing capital improvement plans and identifying specific capital projects for funding tends toward the ad hoc, with utility managers sitting around a table and discussing needs, perhaps using an informal ranking process based on intuition and opinions rather than facts and decision tools. Asset management generates a more rigorous decision-making process based upon quantifiable elements of risk, which ultimately results in more defensible and reproducible capital investment decisions and maintenance practices.

---

[2] Rubenstein, Edwin S., *The Untapped Potential of Water Privatization, A Hudson Institute Report for American Water Works, Inc.,* www.esrresearch.com, October 2006.

[3] GAO-04-461 *Water Infrastructure - Comprehensive Asset Management Has Potential to Help Utilities Better Identify Needs and Plan Future Investments,* March 2004.

[4] *International Infrastructure Management Manual,* Association of Local Government Engineering New Zealand and the Institute of Public Works Engineering of Australia, Version 3.0, 2006.

Better Managed Risk

Because asset management concepts are based on risk evaluation and mitigation, employing asset management principles and practices will inherently allow utility managers to improve their management of risk. If asset risks are scored uniformly across a utility, the risks associated with different assets and different threats can be fairly assessed and risk mitigation can be optimized. For example, comparing the risk posed by the excessive wear of a pump to the risk posed by a malevolent threat on a water storage tank using a uniform risk scoring system can provide utility managers a clear and verifiable process for mitigating risk.

> *Asset management generates a more rigorous decision-making process based upon quantifiable elements of risk...*

Lower Costs

Evidence presented in several case studies included in the GAO report and the *International Infrastructure Management Manual* indicate that asset management can reduce operating and maintenance costs and long-term capital expenses by preserving facility efficiency and avoiding unnecessary investments. In addition, collateral costs from system failures, such as emergency restoration, damage to private property, lawsuits, and fines are minimized or eliminated. Timely rehabilitation extends the life of infrastructure assets and reduces long-term replacement needs. Asset management practices help to identify the optimal point at which an asset should be replaced, thus minimizing overall life-cycle costs.

Improved Public Confidence

Asset management can also improve the relationship utilities have with their customers and external stakeholders and in turn improve the confidence in the utility and the utility's public image. Having factual data about infrastructure assets and being able to clearly communicate service levels, and the asset condition and performance levels required to meet those service levels, can build support from the community, including elected officials and the media.

Improved Bond Rating

Because credit rating agencies often consider criteria related to asset management as evidence the utility is strategically planning its financial future, adopting asset management practices may also improve a utility's bond rating, translating into lower interest rates and lower insurance premiums on new money or refinancing bond issues. The GAO reports, "...according to a representative from one credit rating agency, asset management shows that a utility is considering future costs. He would therefore expect a utility with an asset management plan that looks at future capital and operating costs and revenues to receive a higher bond rating than a utility that does not sufficiently consider those future needs, even if that utility has a better economy and a higher tax base."[5] The *Water and Sewer Revenue Bond Rating Guidelines* published by Fitch, one of the bond rating agencies, state, "Regulations, customer growth, and capacity constraints, as discussed, are each major determinants of a utility's capital improvement burden."[6] In Fitch's view, higher-rated utilities will integrate all of these diverse considerations into a comprehensive, multi-year capital improvement and asset management strategy.

---

[5] GAO-04-461 *Water Utility Asset Management*, 2004.
[6] *Water and Sewer Revenue Bond Rating Guidelines*, Fitch, Inc., 2004.

Possible Higher Priority Ranking for SRF

Bills introduced in Congress reauthorizing funding for state revolving fund (SRF) programs have included a clause giving greater weight in the priority system to a utility that has adopted asset management practices, such as having an inventory of assets that includes a description of asset condition and a schedule for asset replacement. Although this language has not made it into final legislation, it has been introduced several years in a row and may end up being law.

Effective Information Transfer and Knowledge Retention

As addressed in detail in the NACWA and AMWA publications, *The Changing Workforce...Crisis & Opportunity, and The Changing Workforce...Seizing the Opportunity*, a large portion of the existing workforce is approaching retirement and will take with them important institutional knowledge and information about utility infrastructure. The processes and practices of asset management provide the methods and tools to transfer information from employees' experience and expertise and to ensure knowledge is retained within the organization as staffing changes continue over time.

> *The processes and practices of asset management provide the methods and tools to transfer information from employees' experience and expertise and to ensure knowledge is retained within the organization...*

Improved Coordination and Communication

Integrating asset management practices within a utility's operations can facilitate the sharing of information across departments for better coordination and informed decision making. Having readily available, up-to-date, and unambiguous asset information will allow utility managers to reduce duplication of efforts and improve the allocation of staff time and other resources.

Improved Regulatory Compliance

The focus on risk, integration of data, and knowledge of asset condition and other characteristics that are key elements of effective asset management can provide a better understanding of the performance and function of treatment facilities, distribution systems, collection systems, and other utility infrastructure. Reliable and readily available information and data, along with the consequences and likelihood of asset failure, provide the means for improved operations and compliance with permits and regulations.

The benefits of improving the management of utility infrastructure assets through the "new" asset management principles and practices may not be immediately apparent. In fact, it may very well take several years for a utility to fully implement asset management practices in all areas and services. However, if asset management is implemented in the manner presented in this *Practical Guide*, utility staffs will quickly benefit from a more understandable and organized process of programming capital investments and modifying operational and maintenance protocols.

CHAPTER 3

Improved Management of Infrastructure—Asset Management

As mentioned in previous chapters, the management of assets is nothing new for utilities. What is new, however, is the use of fact-based methods, a focus on risk, concepts of level of service combined with life-cycle cost analysis, and the use of integrated tools to improve the ways in which infrastructure assets are managed. This "improved" or "advanced" approach is what is being called asset management.

Asset Management Defined

One of the more succinct definitions of asset management, and the one used in preparing this *Practical Guide*, is based upon the definition published in the 2002 *Handbook* plus the recognition of risk. It defines *asset management* as:

> *An integrated set of processes to minimize the life-cycle costs of infrastructure assets, at an acceptable level of risk, while continuously delivering established levels of service.*

This definition includes four important elements that distinguish today's infrastructure asset management from the manner in which infrastructure assets have traditionally been managed:

- Integrated processes;
- Minimized life-cycle costs;
- Established service levels; and
- Acceptable risk.

First, asset management is recognized to be an *integrated set of processes* rather than a single process or a set of isolated and unrelated activities. While often referred to as a "program," it lacks the defined time period typical of a program; rather, asset management should be a continuous business practice. Asset management processes include investigations, assessments, evaluations, prioritizations, and decision making, among others. Implemented with appropriate decision-making tools, data management, and predictive modeling applications, the processes provide for an iterative system whereby capital investments and operations and maintenance (O&M) protocols are continually refined over time.

Second, the focus is on minimizing life-cycle costs. This includes all costs associated with infrastructure, from planning and design, through O&M, to renewal and eventual replacement and salvage as depicted in Exhibit 3.1.

EXHIBIT 3.1 Components of Life-Cycle Costs

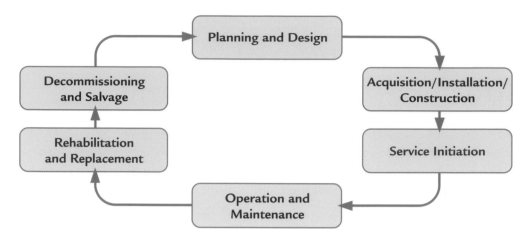

Third, the definition recognizes the purpose for which the utility exists. A utility must establish levels of service to assure the delivery of essential services to its customers, address stakeholder concerns, and provide a safe work environment for its employees. Only with *established* levels of service can the utility know whether or not it is succeeding.

Finally, and very importantly, effective asset management balances the risks between minimizing costs and achieving established service levels, two somewhat contradictory goals. Asset management concepts allow utility managers to quantify and understand the risks posed by failed assets and to manage those risks so that they do not reach unacceptable levels. (A more complete discussion of risk can be found in Chapter 4.)

Key Concepts of Effective Asset Management

Effective asset management can be achieved by focusing on a dozen key concepts. These 12 concepts fall into two categories: (1) Knowledge (awareness, information and data) and (2) Capability (competence, organizational capacity and resources), as shown in Exhibit 3.2.

EXHIBIT 3.2 Key Concepts for Effective Asset Management

Knowledge of:	Levels of service
	Assets and their characteristics
	Physical condition of assets
	Performance of assets
	Total cost of asset ownership
Ability to:	Optimize O&M activities
	Assess asset risk
	Identify and evaluate risk mitigation options
	Prioritize options and fund within available budget
	Predict future demands
	Effectively manage information and employ decision support tools
	Obtain and sustain organizational coordination and commitment

These concepts build upon the approach to asset management contained in the *Handbook*, which was structured along the following basic elements:

· Strategy;

· Asset Retention;

· Tool Integration; and

· Meeting the Challenge and Making the Case.

The *Handbook* presented these elements in a pyramid figure that included the chapter titles above each of the basic elements. Exhibit 3.3 aligns the elements used throughout the *Handbook* with the key concepts addressed in this *Practical Guide*.

EXHIBIT 3.3 **Alignment of the Key Concepts for Effective Asset Management with the Pyramid of Asset Management Elements from the Asset Management Handbook**

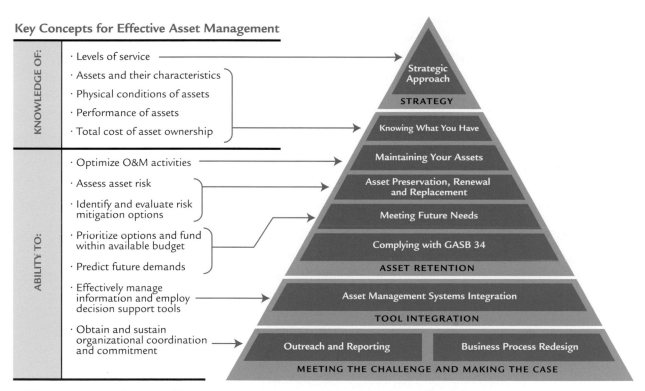

NOTE: *Because most water and wastewater utilities already depreciate their infrastructure assets, they are already in compliance with GASB 34. Thus, this issue is not included as a key concept.*

Each of the key concepts is described below:

Knowledge of...

Knowledge of Levels of Service. This concept examines the extent to which utilities know and measure their current levels of service. Such knowledge is used to determine the condition and performance grade that must be maintained, and thereby enters the decision process for selecting capital rehabilitation and replacement projects, as well as modifying O&M protocols.

Levels of service are established through the development of an asset management strategy that involves the development of an asset management mission and goals, which in turn are based upon the organization's vision and mission. Levels of service typically address the over-arching goals of the asset management mission and represent how infrastructure assets will achieve the goals related to customer service, environmental protection and regulatory compliance, economic sustainability and public and employee health and safety. Alignment with customer desires and expectations is also a key objective in establishing levels of service (Exhibit 3.4).

EXHIBIT 3.4 Alignment Toward Establishing Levels of Service

ALIGNMENT

Levels of Service

Customers & Other Stakeholders

Asset Management Mission & Goals

Organizational Mission & Vision

In countries where asset management has become entrenched in utility business practices (e.g., Australia, New Zealand, the United Kingdom) most levels of service, if not all, are mandated by regulation. This represents an important differentiation in approach, but the intent and outcome of establishing levels of service are the same by reflecting community and environmental goals.

For purposes of a practical approach to asset management, no more than six to eight levels of service should be selected for each utility system (i.e., water system, wastewater system, reclaimed water system). Limiting the number of levels of service allows for a dashboard view of utility performance, and avoids having levels of service that overlap each other, where a failure to meet one level of service automatically impacts the ability to meet another level of service. Exhibit 3.5 lists some desired characteristics to keep in mind when establishing levels of service.

EXHIBIT 3.5 Desired Characteristics of Levels of Service

Meaningful	Relevant to staff and stakeholders Provides a clear picture of performance
Measurable	Can be measured in a cost-effective manner Expressed as a qualitative or quantitative measure
Consistent	Consistent with industry practice Measurement is reproducible by others
Useful	Helps manage the utility Encourages improvement
Unique	Describes a specific attribute of utility services or activities Independent of other levels of service

Knowledge of Assets and their Characteristics. This concept examines the extent to which utilities know what assets they own, the locations of those assets, and information about the assets. It is not necessary to have a complete inventory of all infrastructure assets before implementing asset management. However, information that is available, especially a listing of major assets and possibly groupings of smaller assets that have similar characteristics (called asset groups or asset management units), should be structured in a hierarchy to provide a visual parent-child relationship between the assets. Such a hierarchy can be expanded over time as progress is made with asset management processes. Exhibit 3.6 depicts an asset hierarchy by general terminology and specific assets. Generally, populating an initial asset hierarchy to the fourth or fifth level is a sufficient start.

EXHIBIT 3.6 Asset Hierarchy Terminology

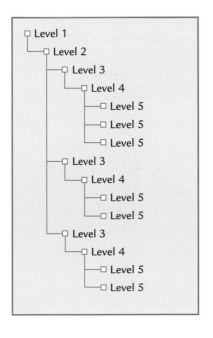
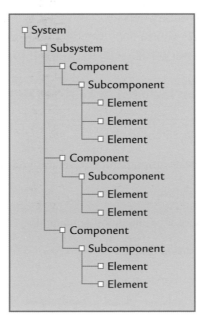
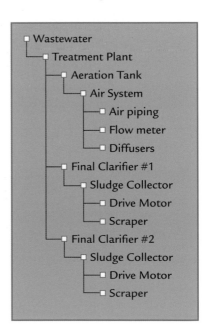

Information about each asset in the hierarchy should be contained in an asset register that is maintained as a database. To ensure consistency of asset information and efficiency of maintaining the asset data, it is preferred to have a single database from which all other software applications draw information. An example of such arrangement is shown in Exhibit 3.7.

EXHIBIT 3.7   Schematic Showing Integration of Asset Register with Other Utility Software Applications

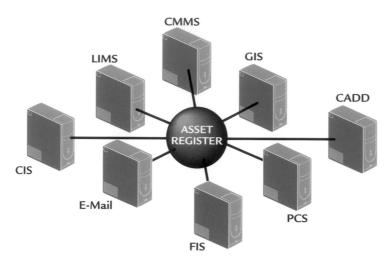

However, if this is not practical, the asset database can be part of other applications such as a computerized maintenance management system (CMMS) for vertical assets (e.g., structures, equipment, etc.) and a geographic information system (GIS) for linear assets (e.g., pipelines). Under such an alternative configuration, all applications should synchronize with the primary asset database to ensure that all data used are consistent and current. Also, each asset should be identified by a unique asset number or identifier code that is linked to the "parent" asset and possibly other descriptors such as location; asset identification systems should be logical and simple. Exhibit 3.8 lists some typical information that is included in an asset register.

EXHIBIT 3.8 Typical Information Included in an Asset Register

Asset ID	Replacement cost
Type of asset	Installation/construction/renewal date
Location	Installation contractor
Manufacturer	Service life (initial and remaining)
Material	Condition grade
Size/capacity	Performance grade
Original cost	Inspection date

Knowledge of Physical Condition of Assets. The physical condition of an asset can have a significant and direct correlation with its likelihood of failure and thereby increase an asset's relative risk (see Chapter 4). An asset's physical condition is typically determined from an in-situ visual inspection and/or from specific tests (which may be part of a predictive maintenance program) such as infrared thermography, ultrasonic measurement, and vibration analysis. There is no right way or wrong way of grading the condition of an asset. However, grading systems typically use a numerical grade from 1 to 5, but in some systems a grade of 1 is the best condition and in others, the worst condition. Some utilities may decide to use a relatively straight-forward grading system that can be used for all assets, as shown in Exhibit 3.9.

EXHIBIT 3.9 Basic Asset Condition Grading System

GRADE	DESCRIPTION OF CONDITION
1	**Very Good Condition** Only normal maintenance required.
2	**Minor Defects Only** Minor maintenance required—5% of the entity needs maintenance.
3	**Backlog Maintenance Required to Return to Accepted Level of Service** Significant maintenance required—10 to 20% of the asset needs maintenance.
4	**Requires Renewal** Significant renewal/upgrade—20-40% of the asset needs renewal.
5	**Entity Unserviceable** Over 50% of the entity requires replacement.

Other utilities may decide to use grading systems more specific to asset type. For example, several condition scoring systems are available for gravity sewers (e.g., NASSCO, WRc, NRC-Canada, etc.). Additionally, the New Zealand Water and Wastes Association publishes grading guidelines[7] that are specific to numerous categories of infrastructure assets. Regardless of the grading system that is used, it must be employed consistently across the set of assets for which it was developed.

Knowledge of Asset Performance. Often considered jointly with physical condition, performance is an asset's ability to meet its intended purpose, whereas condition is the physical state or structural integrity of an asset. Asset performance and asset condition are linked in that the performance of an asset can be affected by its condition. However, an asset can be in excellent condition but still be performing poorly if it has insufficient capacity or is inappropriate for the application. For some assets, performance can be monitored continuously through supervisory control and data acquisition (SCADA) systems or be evaluated on a periodic basis through measurements. Performance can be graded on a scale similar to that of condition, or it may be assessed by the specific issues affecting performance, such as capacity, utilization, functionality, and delivery of desired level of service.

---

[7] Adapted from *New Zealand Infrastructure Asset Grading Guidelines–Water Assets*, New Zealand Water and Wastes Association, 1999.

Knowledge of Total Cost of Asset Ownership. To minimize life-cycle costs, the total cost of asset ownership must be known. The cost categories, which were depicted in Exhibit 3.1, earlier in this chapter, are:

- Planning and design;

- Acquisition, installation, and construction;

- Service initiation;

- Operation and maintenance;

- Rehabilitation and replacement; and

- Decommissioning and salvage.

It is important for these costs to be captured by asset, or asset group, throughout an asset's life-cycle. This can be accomplished by careful accounting and allocation of labor, material and utility costs associated with each asset or asset group. It can also be achieved through activity-based costing that tracks costs by activities performed in the production of a product (in this case drinking water, effluent, or reclaimed water).

Ability to...

Ability to Optimize O&M Activities. Frequently, because of the relatively low cost of implementation, the most effective and efficient risk reduction options will be refinement of existing, or adoption of new, O&M protocols. Such protocols should be based upon optimizing the mix between planned maintenance (i.e., preventive maintenance and predictive maintenance), and unplanned (reactive) maintenance. As described in the *Handbook*, the generally accepted mix of maintenance activities to achieve the lowest total maintenance cost is 70 to 80 percent planned and 20 to 30 percent unplanned. Additional guidance on optimizing O&M processes can be found in the NACWA and AMWA publication, *Thinking, Getting, and Staying Competitive: A Public Sector Handbook*.[8]

Ability to Assess Asset Risk. Being able to identify, quantify, and assess the risks posed by asset failure is the key asset management concept. By determining the consequence of an asset failure and the likelihood of an asset failure, the risk of asset failure can be calculated. Assets can then be ranked by relative risk, and those having the highest risks can receive the attention of the utility, and appropriate risk mitigation options, such as capital renewal projects or operational enhancements, can be put into place. Because the concept of risk is so important to asset management, Chapter 4 is dedicated to this subject.

Ability to Identify and Evaluate Risk Mitigation Options. Where the risk of asset failure is determined to be greater than acceptable, options for reducing the risk to acceptable levels must be identified and evaluated. Risk mitigation options may include capital improvement projects, changes in operational or maintenance protocols, reduction of service levels (as long as regulatory requirements continue to be met), demand management, or improved response and recovery mechanisms. Identified options should be evaluated for life-cycle costs, risk reduction, and their ability to be effectively implemented.

---

[8] *Thinking, Getting, and Staying Competitive: A Public Sector Handbook*, Association of Metropolitan Sewerage Agencies and Association of Metropolitan Water Agencies, 1998.

Ability to Prioritize Options and Fund within Available Budget. In any utility it is likely that there will be numerous risk mitigation options to address several high risk assets. Yet, there is always a limited budget, whether it is capital or O&M, and some risk reduction actions will necessarily be deferred. Being able to prioritize options to determine the projects or O&M activities that should be funded first and those that should be deferred for subsequent years can be accomplished through a risk-reduction vs. cost analysis. In such an analysis, the risk reduction for each option is calculated along with the cost of each option. In general, those options having the greatest risk reduction per dollar are given the highest priority and funded first. Further discussion on prioritizing risk reduction options is presented in Chapter 4.

> *Frequently,...because of the relatively low cost of implementation, the most effective and efficient risk reduction options will be refinement of existing, or adoption of new, O&M protocols.*

Ability to Predict Future Demands. This concept is critical in determining an asset's ability to meet future capacity needs. It also helps a utility to determine which assets need to be modified or augmented with respect to function, capability, or capacity, and where future investment may be needed. Forecasting methods can be either qualitative or quantitative. Qualitative methods are most applicable to predicting the impact of future regulations and customer expectations; quantitative methods, usually involving computer models, are typically used to forecast capacity demands and cost impacts.

Ability to Effectively Manage Information and Employ Decision Support Tools. The enormous volume of asset data and its efficient use to improve operations, maintenance, and capital investment decisions requires the use of information systems and decision support tools. With the wide variety of asset management-related tools available, even smaller to mid-size utilities can benefit from implementing business-driven technology solutions to better manage their assets, rather than relying on employees' memory and paper records. Important principles of employing information technology include the following:

· Ensure accessibility and usability of applications and data;

· Minimize the number of applications and data sources;

· Develop and implement an enterprise architecture that is followed by all key utility stakeholders;

· Use commercial off-the-shelf (COTS) software applications when and where possible to lower the life-cycle costs of IT infrastructure;

· Use a single database for the asset register (when and where possible) that links to other applications;

· Create data protocols and reports based on actual work practices; and

· Maintain a secure network and protect data.

More information on information technologies is presented in Chapter 6.

Ability to Obtain and Sustain Organizational Coordination and Commitment. Although asset management does not require an entire organizational cultural change, its success is dependent upon the adoption of new business processes and some change management, especially with respect to decision making and data integration. Consequently, it is imperative to assure all staff members are aware of the asset management mission and the processes being implemented to improve the manner in which information about infrastructure assets is gathered and analyzed, and how that information is used to enhance the capital planning process and optimize O&M activities.

Although asset management does not require an entire organizational cultural change, its success is dependent upon the adoption of new business processes and some change management, especially with respect to decision making and data integration.

CHAPTER 4

Focusing on Risk

"Risk" may be the most important concept of asset management. By quantifying and assessing the risks posed by the failure or inability of infrastructure assets to meet their intended functions or achieve overall levels of service, utility staff can identify operating and maintenance procedures, as well as capital rehabilitation and replacement projects, to mitigate the risks. By determining the level of risk reduction for each risk mitigation option along with its associated cost, the options can be prioritized based upon a comparison of the ratio of the level of risk reduction to the cost for each option identified.

As described in Chapter 3, effective asset management is a balance of minimizing life-cycle costs while continuously providing the levels of service established by the utility to meet customer and stakeholder expectations. Of course, it is rather simple to minimize costs if one is not concerned about maintaining given service levels. Likewise, it is rather simple to meet established service levels if cost is not a factor. However, the challenge to utilities is that they must do both concurrently. This is where risk must be managed—the risk resulting from too little investment in infrastructure assets and not meeting established levels of service and the risk of overspending to offer levels of service above and beyond what may be required or expected by customers and stakeholders.

> *"Risk" may be the most important concept of asset management.*

Quantifying Risk

In everyday use, "risk" is a concept that relates to the expectation of a negative impact generated by some action or inaction. Commonly, "risk" is used synonymously with the likelihood (or probability) of a negative impact occurring. Sometimes "risk" is used to describe severity of the consequence of a potential failure. However, it is the combination of both of these factors, likelihood **and** consequence, that contributes to risk.

Mathematically, it is widely accepted that risk from an incident or occurrence can be expressed as a function of the consequence of the incidence or occurrence and the likelihood of the incidence or occurrence. Thus, risk can be quantified through the equation shown in Exhibit 4.1.

EXHIBIT 4.1 Mathematical Expression for Risk

Risk equation: Risk = [(Consequence) x (Likelihood)]

For the purposes of infrastructure asset management, the risk is the risk of asset failure, the consequence is the impact on established levels of service, and the likelihood is the possibility of asset failure.

The risks posed by the failure of infrastructure assets are not only the result of physical failure of the asset. Asset risk also results from an asset's inability to perform to its function or purpose, such as providing sufficient capacity or maintaining a safe environment for employees. Additionally, risk can be attributed to indirect consequences of an asset's physical failure or its failure to perform, such as payment of damage claims, litigation costs, and higher insurance premiums. Therefore, when quantifying risk through an evaluation of its inputs (i.e., consequence and likelihood), it is important to consider the consequences from the broader idea of an asset being unable to meet its purpose rather than just a physical or functional failure. Similarly, the second input to the risk equation, likelihood, should be based on an asset's overall inability to meet its purpose rather than just the likelihood of a physical or functional failure.[9]

To quantify risk using the risk equation, the two variables, consequence and likelihood, must be defined and quantified. Consequence is the resulting outcome when an asset physically fails or fails to meet its purpose. Identification of consequences should be targeted to the utility's levels of service. When evaluating consequence, the effect of asset failure on each of the utility's levels of service should be assessed. Therefore, before embarking on an analysis of consequence, the utility's levels of service must be established. (See Chapter 3 for discussion on levels of service.)

For the purposes of quantifying the consequence of asset failure, it is useful to develop a matrix that lists the utility's level of service (LOS) down the far left hand column and the consequence effect and score across the top row of the matrix. Each cell can then be filled in with a description of the consequence by category and by effect. Also, each category may be given a weighting based on the utility's assessment of which category (i.e., level of service) may be more important to the organization and its stakeholders. An example of this matrix is shown in Exhibit 4.2. A more detailed discussion of consequence matrices is contained in Chapter 5.

EXHIBIT 4.2 Example Consequence Matrix

Consequence by LOS Category						
			Consequence			
LOS Category	Weight	Negligible = 1	Low = 4	Moderate = 7	Severe = 10	
Category 1						
Category 2						
Category 3						
Category 4						
Category 5						
Category 6						

---

[9] It is important to note, that an assessment of functional failures through rigorous methodologies such as reliability-centered maintenance (RCM) is appropriate for some high-risk assets. However, a description of such methodologies is beyond the scope of this *Practical Guide*.

Likelihood is the possibility that an asset will physically fail or otherwise fail to meet its purpose (i.e., established levels of service). Likelihood reflects the characteristics of an asset and its care. Likelihood categories can include an asset's physical condition, its ability to meet capacity needs, and the effectiveness of operational and maintenance protocols.

To quantify likelihood, as with consequence, it is useful to develop a matrix. The categories that affect an asset's likelihood of failure are also listed down the far left column and the range of likelihoods across the top row, each with an associated likelihood score. The cells can then be filled in with a description of situations that contribute to the range of likelihoods. Also, each category of likelihood may be given a weighting based on the utility's assessment of how the category may contribute more to the asset's likelihood of failure. An example of a likelihood matrix is shown in Exhibit 4.3. A more detailed discussion of likelihood matrices is contained in Chapter 5.

EXHIBIT 4.3 Example Likelihood Matrix

Likelihood Category						
		Likelihood				
Likelihood Category	Weight	Negligible = 1	Unlikely = 2	Possible = 4	Likely = 7	Very Likely=10
Category 1						
Category 2						
Category 3						
Category 4						
Category 5						

The categories, assigned weights, and scoring systems for the consequence and likelihood matrices should be developed with consistency in mind. To equitably assess risk across the utility, the matrices must be applicable to all assets. While a water system, wastewater system, and reclaimed water system will have matrices focused on their particular characteristics, consequence and likelihood categories should be the same for all systems, as should the scoring system. Wherever practical, category weights for the different systems should be the same, as should the majority of the definitions in the cells of the matrices. The closer the matrices of the different systems match, the more valid the risk comparisons will be between the utility's assets.

Applying the risk equation to each of the assets under consideration will allow for a comparison of assets by their relative risks. Because the scoring is somewhat subjective, the absolute risk value is not as important as is the risk rank of an asset. By listing assets by their risk value, it is usually clear which assets are posing the highest risk, which are posing the next highest risk, and so on. Risk grades can then be given by groupings. For example, by calculating the difference in risk score from one asset to the other, a clear distinction can usually be made where one grade grouping should stop and another begin. One way of displaying the results is by grading the assets by risk, as shown in Exhibit 4.4. In this case, asset grades were changed whenever the risk score between assets was 9 or greater.

EXHIBIT 4.4   Example of Asset Risk Grading

Selected System Assets			
ASSET	Risk Score	Risk Score Δ	Risk Grade
Asset 2	78	10	E
Asset 10	68	4	D
Asset 7	64	1	D
Asset 16	63	1	D
Asset 8	62	1	D
Asset 6	61	12	D
Asset 12	49	0	C
Asset 4	49	3	C
Asset 14	47	1	C
Asset 15	46	8	C
Asset 13	38	4	B
Asset 5	34	2	B
Asset 1	32	9	B
Asset 11	23	3	A
Asset 9	20	4	A
Asset 3	16	—	A

Risk Mitigation

For those assets considered to have a higher than acceptable risk score or unacceptable risk grade, a review of the consequence and likelihood scoring will provide insight into the issues that resulted in the relatively high risk of the assets. Utility staff can then identify risk mitigation options that can be assessed to lower the assets' risk. Asset risk can be mitigated either by reducing the consequence of failure or the likelihood of failure, or both, as depicted in the risk matrix shown in Exhibit 4.5.

EXHIBIT 4.5 Risk Matrix Showing Alternative Methods for Reducing Risk

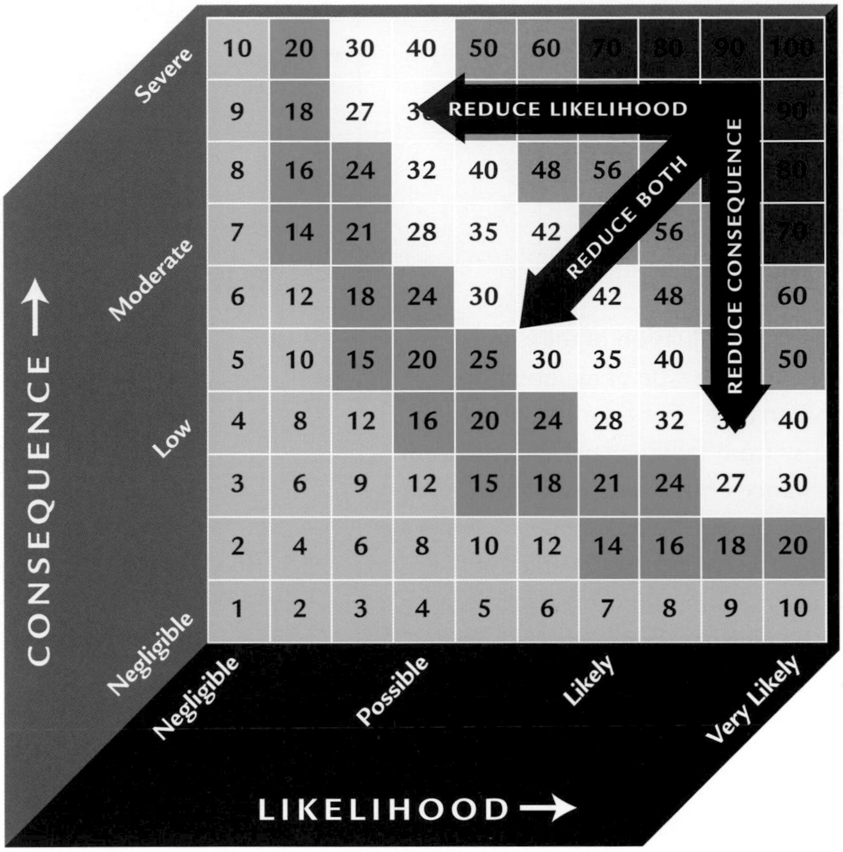

Typically, capital rehabilitation and replacement projects are the first risk mitigation options identified. However, modifications to O&M practices can also reduce risk, and usually at lower cost. Additionally, some less obvious means of reducing risk include (1) demand management, (2) reduction in levels of service[10], and (3) improved response and recovery.

One method for evaluating and prioritizing risk mitigation options is to estimate their cost and the amount of risk reduction they will provide. Those options having the highest risk reduction to cost ratio may be considered as having the highest priority.

The numerator in the ratio, *risk reduction*, is the difference between (1) the risk score of the existing asset determined using the matrices and risk equation, and (2) the risk score of the asset assuming the option was implemented (i.e., re-scoring the risk with the matrices assuming the asset was rehabilitated or replaced, some O&M protocol was changed, etc). The denominator in the ratio is the cost of the risk reduction option (i.e., cost of rehabilitation or replacement, or cost of implementing and practicing changed O&M protocols). The cost should be calculated as a present value of both initial and recurring costs. The present value calculation for all options should have the same term and rate to ensure equitable comparisons.

---

[10] Any proposed reduction in service levels should be carefully vetted with the utility's governing body, customers, and other stakeholders before adoption.

Exhibit 4.6 shows an example of how several capital improvement projects to reduce an asset risk might be assessed. CIP projects #3 and #6 have the highest *risk reduction to cost ratio* and therefore may be considered preferred options, and eligible for further evaluation. Chapter 5 shows an example of how the risk reduction to cost ratio is applied.

Asset Risk and Asset Maintenance

Utilities can also use the evaluation of the risk from failure to help determine the type of maintenance to be performed on an asset. For example, an asset posing a very low risk from failure may be considered for reactive maintenance only. Assets with a medium risk may be slated for preventive maintenance, while higher risk assets may be targeted for predictive (i.e., condition-based) maintenance. For those relatively few assets posing the highest risk to the utility, the more intensive (and expensive) maintenance approach of reliability-centered maintenance (RCM) may be appropriate. Studies[11] have shown that, although it can be costly to establish a preventive maintenance program, the estimated savings in maintenance as compared to a reactive maintenance can be from 12 to 18 percent or greater. Likewise, going from preventive maintenance to predictive maintenance is also costly, but estimated savings in maintenance range from 8 to 12 percent or more.

EXHIBIT 4.6 Example of Assessing CIP Risk Mitigation Options

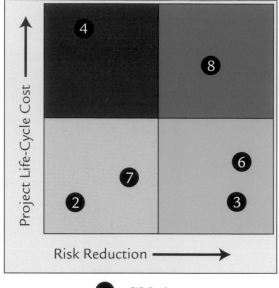

● = CIP Project

Asset Redundancy and Asset Risk

There are some different approaches to addressing asset redundancy in calculating the risk of asset failure. Some consider redundancy to be a category in a likelihood matrix and score likelihood high for assets that have no redundancy and score likelihood low for assets with multiple redundancies. Others consider redundancy to be a category in a consequence matrix and score the consequence higher if there is no redundancy for the asset and low if there are other assets that provide redundancy for the asset being evaluated. However, redundancy is not a characteristic of an asset and does not affect its likelihood of failure. Therefore, redundancy should not be factored into the likelihood variable of the risk equation or in a likelihood matrix.

Redundancy does affect the consequence of asset failure, but is not a unique consideration. Consequently, redundancy should be considered when evaluating the consequences of asset failure, but because it affects more than one consequence category, it should not be a separate consequence category. For example, when assessing the risk of failure of a pump at a pump station, having a redundant pump can (1) ensure uninterrupted service to customers, (2) maintain regulatory compliance, and (3) minimize the financial impact on the utility. Redundancy is discussed further in Chapter 5.

---

[11] *Operations & Maintenance Best Practices A Guide to Achieving Operational Efficiency*, Release 2.0. Pacific Northwest National Laboratory for the Federal Energy Management Program U.S. Department of Energy, July 2004.

CHAPTER 5

Implementing Asset Management

Probably the most daunting challenge faced by utility mangers in implementing asset management is deciding where to begin. The sheer volume of infrastructure assets in a utility can quickly overwhelm the staff to the point of giving up on implementing asset management.

Overcoming this challenge and reducing the apprehension that likely accompanies the implementation of asset management is best accomplished through an understanding of the two basic approaches to asset management and the way to leverage the advantages of each approach while minimizing the impact of their disadvantages.

> *Probably the most daunting challenge faced by utility mangers in implementing asset management is deciding where to begin.*

Approaches to Implementation

There are two basic approaches to implementing asset management:

1. The Bottom-Up Approach

2. The Top-Down Approach

The **Bottom-Up Approach** is the method most utility professionals think about when embarking on advanced asset management processes. It involves collection of detailed characteristics of all the utility's assets (e.g., type, material, manufacturer, size, capacity, etc.), a review of operational and maintenance records, field condition assessments, performance analysis, estimation of remaining useful life, determination of asset replacement costs, and other detailed information. All of the asset attribute information collected is then entered into a database to create an asset register (see Chapter 3). The risks posed by the assets are assessed; risk mitigation measures are identified and selected.

Because of the detailed investigations undertaken through the Bottom-Up Approach, the data collected are of relatively high quality and can be used with a high degree of confidence when employing predictive modeling and making decisions to mitigate asset risks. However, the Bottom-Up Approach requires a considerable financial investment, is resource intensive from both the involvement of personnel and equipment, and because of all the data collection required there is a significant time lapse from beginning the approach until results are achieved that are valuable to the organization. Thus, there is a potential for the utility to become data-rich but information-poor and for the utility staff to lose focus by not realizing quicker benefits.

The **Top-Down Approach**, on the other hand, focuses first on a group or set of assets at a system or facility level. It makes the most of existing and available data, as well as institutional knowledge, to assess consequence, likelihood and risk, and to identify and select risk mitigation actions.

Because the Top-Down Approach does not involve detailed field investigations, hands-on condition assessments, or sophisticated performance evaluation, the human and equipment resources required are considerably less than those needed for the Bottom-Up Approach. Also, estimates of asset risks can be determined relatively quickly, resulting in valuable and timely information for the minimal investment of resources. However, because the Top-Down Approach does not involve gathering detailed asset information, numerous assumptions and projections are usually required. This creates the potential for incomplete or inaccurate data and possibly lower confidence in the decisions that are made based on the Top-Down Approach.

Because both approaches have advantages and disadvantages, the most effective method of implementing asset management is to utilize the best elements of each approach in a manner that provides reliable decision making in a relatively short term while optimizing the financial investment and organizational resources. This can best be accomplished by combining the two methods in an iterative fashion. Beginning with the Top-Down Approach yields relatively quick and valuable information that can be used to improve capital renewal and O&M decisions. It also provides the direction as to where investments and resources should be focused during the Bottom-Up Approach to gather detailed asset information and data that can refine decision making. As depicted in Exhibit 5.1, this combination of the Top-Down Approach with the Bottom-Up Approach provides a utility with continuous improvement in managing its assets over time.

EXHIBIT 5.1 Combining the Top-Down and Bottom-Up Approaches

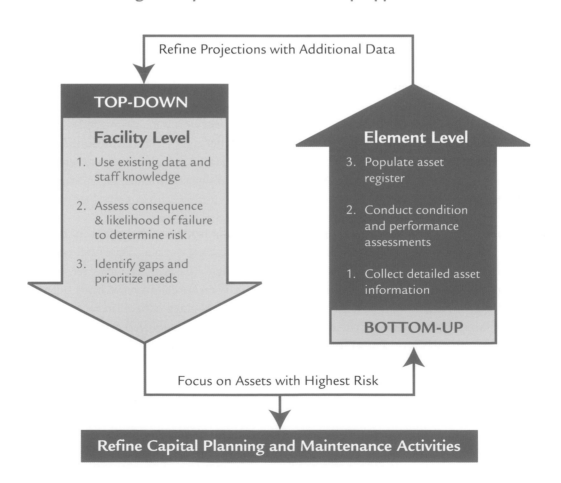

Implementation Overview

A logical means for implementing asset management is with the Top-Down approach applied first at the higher levels of the asset hierarchy. Assessing risk at the highest levels of a water, wastewater, or reclaimed water system allows for the determination of relative risk among the different facilities and guides the utility as it drills down to the next level of the hierarchy. By proceeding in this manner to the fourth or fifth level of an asset hierarchy, in the order of highest risk assets to lower risk assets, the following benefits can be quickly achieved:

1. Assurance that the utility is addressing the highest risk assets through either its capital improvements plan or changes to O&M practices.

2. Identification of projects, O&M practices, and management actions that can result in a more effective reduction in asset risk.

3. Focus resources on the assets at highest risk to obtain more detailed asset information, assess physical condition, and evaluate asset performance.

The implementation begins by setting the foundation of asset management, as follows:

·   Establishing the asset management strategy aligned with the organization's mission and goals;

·   Establishing levels of service;

·   Developing the asset hierarchy;

·   Preparing matrices and scoring systems (used to calculate relative risk of asset failure).

The matrices and scoring systems are applied first to the assets at the highest level of the hierarchy to determine risk, risk-reduction options, and focus areas for collecting more detailed information. The same is done for the next level of assets down to the fourth or fifth level of the hierarchy. As more detailed

This is a continuous, iterative process...

information is gathered, the process is repeated and the analysis is refined based on the more detailed information. This is a continuous, iterative process that, when diligently performed, will provide a utility with the best information on which to minimize life-cycle costs while delivering established levels of service at an acceptable level of risk. Exhibit 5.2 is a flow diagram of asset management implementation; it is followed by a description of each step in the process.

EXHIBIT 5.2   Asset Management Implementation Flow Diagram

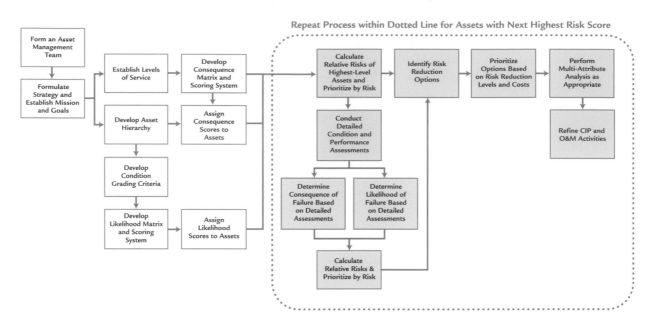

Form an Asset Management Team

To assure development of practical and sustainable asset management, it is imperative to involve utility staff from the beginning. This is best accomplished through the establishment of an Asset Management Team that will serve as the core leadership for implementing asset management and maintaining its progress. The team should include staff members from all of the utility's business units or departments, including operations, maintenance, engineering, planning, finance, customer information, government and regulatory relations, purchasing, and any other parts of the organization that influence capital investment and O&M decisions. Exhibit 5.3 shows some examples of Asset Management Team members for small, medium, and large utilities.

The Asset Management Team is responsible for developing the asset management foundation, including the strategy, levels of service, matrices, and scoring systems. In developing the foundation materials, the Asset Management Team should work interactively to minimize individual perceptions becoming fact and ensure decision making through consensus. The individual members of the Asset Management Team should create functional teams consisting of staff members from the Asset Management Team members' respective functional areas. These functional teams should be available to assist in gathering specific, existing information about assets needed to perfect the asset hierarchy, data, and knowledge to score assets for consequence and likelihood.

EXHIBIT 5.3 Example Asset Management Team Members

Small-Size Utility	
Utilities Director	Budget Analyst
Treatment Superintendent	Chief Plant Operator(s)
Field Superintendent	Maintenance Foreman
Capital Projects Coordinator	Analyst from City/County IT Department

Medium-Size Utility	
Deputy Utilities Director	Water Distribution Supervisor
Treatment Operations Manager	Wastewater Collections Supervisor
Wastewater Plant Manager	Engineering Manager
Maintenance Manager	Utilities Finance Manager
Field Operations Manager	Utilities Database Administrator

Large-Size Utility	
Deputy Executive Director	Maintenance Manager
Director of Water	Field Superintendent(s)
Director of Wastewater	Director of Engineering
Water Production Manager	Director of Planning
Wastewater Treatment Manager	Director of Finance
Distribution System Manager	Director of Customer Service
Collections System Manager	Director of Information Technologies
Treatment Plant Manager(s)	Public Information Officer

Formulate Strategy and Establish Mission and Goals

For asset management to be successful and sustainable, it must be aligned with the organization's overall vision and mission.[12] The levels of service that will form the basis for evaluating the consequences of asset failure should be developed from the utility's strategic mission and goals. Thus, it is important to begin with a review of the utility's mission and vision statements. It is likely that many, if not all, of the key concepts contained in the organization's mission and vision statements will be applicable to asset management. The Asset Management Team should craft an asset management mission statement and goals using these same concepts to guide the establishment of good asset management practices, including development of realistic and measurable levels of service.

Utilities do not need to start this project from scratch. Many asset management mission statements and goals are available to use as a starting point, as shown in Exhibits 5.4 and 5.5.

> *For asset management to be successful and sustainable, it must be aligned with the organization's overall vision and mission.*

---

[12] If a utility does not have mission and vision statements, its leadership should consider convening a strategic planning effort with its management to formally establish its purpose (mission) and its image of the future (vision) before implementing asset management.

EXHIBIT 5.4 Example Asset Management Mission Statements

- To ensure that all assets are managed in a way that provides the desired level of service in the most cost-effective and achievable manner for existing and future customers.

- To attain a leadership role in the water and wastewater industry through the provision of excellent services that will meet the needs of customers and enhance the value of the utility to its stakeholders.

- To meet established levels of service in the most cost-effective way through the creation, operation, maintenance, renewal, and disposal of assets for existing and future customers.

- To maximize the public's return on its investment in utility infrastructure by implementing utility-wide strategies that emphasize reliability in the assets and processes so that the desired levels of service are provided to our customers in the most cost-effective manner.

- To maintain, upgrade, and operate physical assets cost-effectively through a systematic process that combines engineering principles with sound business practices and economic theory.

- To improve the condition of infrastructure assets through cost-effective renewal and replacement investments that minimize the overall costs to customers over the life cycle of the assets.

- To develop funding and supporting policies that improve the overall condition of the assets over time to a level that reduces reliance on costly emergency repair of failing assets.

- To employ a holistic approach for decision making, investment analysis, and management of our infrastructure assets through a process that links policies with financial planning, programming, and performance monitoring to continuously improve the efficiency and effectiveness of our assets.

- To optimize the preservation, upgrade and timely replacement of assets through cost-effective management, programming, and resource allocation decisions.

EXHIBIT 5.5 Example Asset Management Goals

- To monitor changes in the service potential of assets and identify deferred maintenance.

- To manage environmental and financial risks associated with asset failure.

- To achieve savings by optimizing life-cycle work activity.

- To maintain a ratio of system deficiencies to replacement value that remains the same or decreases over time.

- To maintain or enhance the capability and integrity of assets to meet growing and changing demands in a cost-effective manner.

- To demonstrate responsible stewardship of the network assets.

- To collect data on physical plant inventory and manage current conditions.

- To maintain an up-to-date electronic inventory of infrastructure assets that includes acquisition and cost information.

- To maintain ongoing maintenance and repair logs that track the cost of maintaining strategic assets.

- To monitor advances in the (water or wastewater) industry to facilitate continuous improvements in the asset management program.

- To provide a basis for customer consultation over the price/quality trade-offs resulting from various levels of service.

- To renew and replace assets in a cost-efficient manner that maximizes opportunities to reduce costs through strategic intervention where such intervention will reduce overall life-cycle costs when compared with running the assets to failure.

Establish Levels of Service

The terms *levels of service* and *performance measures* are often used interchangeably. However, for the purposes of implementing asset management, levels of service should be defined as those objectives that are established at a utility-wide level, while performance measures should be established at lower levels of the organization to measure and track performance of functions and relationships between inputs and outputs. Levels of service, like performance measures, should be established with targets, but the targets for levels of service may be qualitative as well as quantitative, whereas performance measures have quantifiable targets. As discussed in Chapter 3, no more than six to eight levels of service should be selected to maintain a manageable dashboard view of utility performance and to ensure uniqueness and independence between each chosen level of service.

The Asset Management Team should review any existing levels of service already adopted by the utility for consistency with the asset management mission statement. Existing levels of service should be revised, and new levels of service created, with organizational alignment in mind. Also, levels of service should not be established in the vacuum of a meeting room. Team members must understand and take into consideration the issues and expectations deemed important and relevant by their customers and other stakeholders, including utility staff. Some examples of level of service categories and targets are shown in Exhibit 5.6.

EXHIBIT 5.6   Example Levels of Service

	Level of Service	Target
Water	Compliance with primary drinking water standards	100%
	Pressure at customer side of meter	35 psi minimum
	Restoration of service from unplanned outage	4 hours maximum
Wastewater	Compliance with effluent discharge permits	100%
	Sanitary sewer overflows	Zero
	Odor complaints	2 per month maximum
General	Employee health and safety	Zero injuries
	Competitive rates	Within 5% of surrounding utilities
	Public image	No adverse media reports

Develop Asset Hierarchy

An asset hierarchy is a representation of the relationships between a utility's infrastructure assets. It is arranged as a family tree in a parent-child format; that is, the hierarchy depicts the relationship of an asset to another asset, starting at the top with the utility system, then to the facility, followed by components of the facility, etc. Asset hierarchies may be configured by physical location or by function; neither is right nor wrong, and either will likely pose some challenges along the way. If a utility already has a computerized maintenance management system (CMMS), consideration should be given to arranging the asset hierarchy in the same manner as it exists in the CMMS. Exhibit 5.7 shows two of the many ways that an asset hierarchy may be arranged.

With the assistance of the functional teams, the Asset Management Team should develop asset hierarchies for each utility system (water, wastewater, reclaimed water) as appropriate. Beginning with the Top-Down Approach, it will be necessary to populate the hierarchy only to the fourth (subcomponent) or fifth (element) level. Assets that would be listed at the lower levels should be grouped into their higher level parent asset for evaluation.

EXHIBIT 5.7   Example Asset Hierarchies

Partial Hierarchy: Example 1

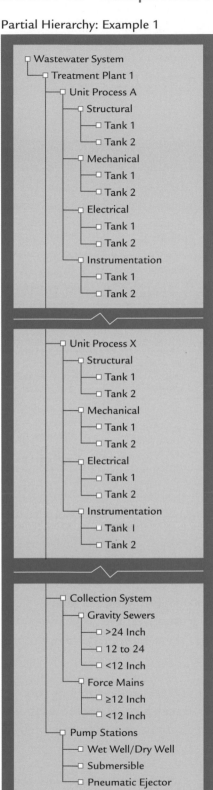

Partial Hierarchy: Example 2

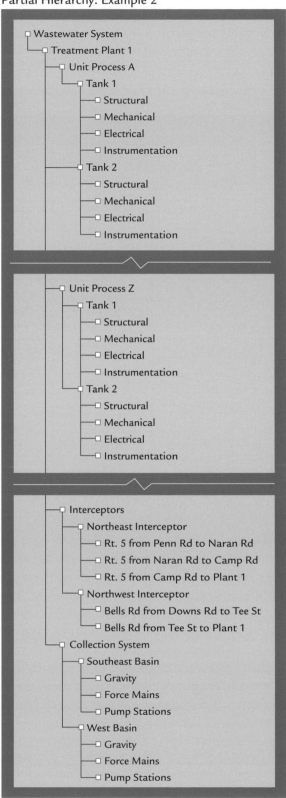

Develop Consequence Matrix and Scoring System

As discussed in Chapter 4, it is suggested that the consequence variable of the risk equation be quantified by the use of a consequence matrix. This matrix provides guidance to the asset evaluators by defining different consequences resulting from asset failure and provides a consequence score for the severity of the consequence. An example consequence matrix is shown in Exhibit 5.8.

EXHIBIT 5.8   Example Asset Failure Consequence Matrix

Consequence by LOS Category					
Consequence Category	Weight	Negligible = 1	Low = 4	Moderate = 7	Severe = 10
Health & Safety	0.20	No injuries or adverse health effects	No lost-time injuries or medical attention	Lost-time injury or medical attention	Loss of life
Compliance with Regulations	0.20	100% compliance with permits	Technical violation but no enforcement action	Violation with minor enforcement action	Enforcement action with fines
Financial Impact	0.10	Absorbed within budget line item	Absorbed within current budget	May require transfer from reserves	May require new borrowing or impact rates
Disruption to the Community	0.15	No social or economic impact	Minor disruption (e.g., traffic, dust, noise)	Short-term impact; substantial disruption	Long-term impact; area-wide disruption
Service Delivery	0.20	No overflows, backups, or odors	No dry weather overflows or backups; infrequent odors	Short duration dry weather overflows or backups; occasional odor	Numerous overflows, backups; widespread or persistent odors
Ability to Respond and Continue Service	0.15	< 2 hours	2 to < 8 hours	8 ≤ 24 hours	> 24 hours

The established levels of service are the consequence categories against which assets are evaluated. These are listed in the far left column. The categories may be weighted so that more important levels of service categories have a greater effect on the consequence score. The weights should be determined by consensus of the Asset Management Team. It is suggested that the weights be normalized (i.e., the total of the individual weights add to 1.0) so that the highest possible risk score is 100; however, normalized weightings are not required.

The levels of severity listed across the top row (e.g., negligible, low, moderate, and severe) can be chosen by the Asset Management Team. The score (e.g., 1, 4, 7 and 10) can also be chosen by the Team. However, it is suggested that the 1 to 10 scale be maintained to avoid a concentration of final scores.

The cells in the first column, titled "Negligible" in the example above, should be populated with the level of service targets. That is, should an asset fail and the consequence does not cause a breach of level of service target, then the consequence is the lowest possible, or negligible.

The cells in the remaining columns should be completed by the Asset Management Team based on the asset management mission, organizational policies, understanding of customer and other stakeholder expectations, and institutional knowledge.

Develop Condition Grading Criteria

The physical condition of an asset is a dominant factor in determining its likelihood of failure. As discussed in Chapter 3, condition grading systems can be simple or complex. Some infrastructure condition grading systems incorporate scores from 1 to 100, others from 0 to 10 or 1 to 5, or some other numerical range. Not only is there a lack of uniformity in the ranges of grading systems, there is also no consistency in whether the lower or higher number represents the better condition.

While uniformity among different infrastructure condition grading systems has not yet been achieved, uniformity within a utility is critical. Further, the grading system chosen should allow for simple mathematical manipulation of the numerical grade to properly reflect likelihood of failure and asset risk. Therefore, it is recommended that utilities that have not yet established a system for condition grading select a condition grading system from 1 to 5 where 1 is the best condition and 5 is the worst. In this system, the condition grade directly reflects the likelihood of failure and asset risk; the higher the grade or score, the more negative impact on the utility.

The criteria used to determine the condition grade can be of a general nature, such as describing the asset's overall maintenance need, or can be specific to the type of asset by describing distinct characteristics of an asset's condition, such as sewer defect codes developed by NASSCO, WRc, and others. While describing distinct characteristics of an asset's condition may provide more reproducible results among different condition assessors, it requires too much detailed inspection for the Top-Down Approach. Consequently, the adoption of a generic condition grading system is suggested. An example of such a system was presented in Chapter 3.

Develop Likelihood Matrix and Scoring System

The Asset Management Team should develop a matrix similar to the consequence matrix to provide guidance in the scoring of the likelihood of asset failure. The categories to be scored in this matrix should include physical condition and functional performance, as well as other criteria that may reflect on the likelihood of an asset failure as determined by the Team. These categories are listed down the far-left column of the matrix. The categories may be weighted so that more important categories (e.g., physical asset condition) have a greater effect on the consequence score. The weights should be determined by consensus of the Asset Management Team. As with the consequence matrix, the Asset Management Team may want to consider normalizing the weights so that the total of the individual weights adds up to 1.0.

An example of a likelihood matrix is shown in Exhibit 5.9.

EXHIBIT 5.9 Example Asset Failure Likelihood Matrix

		Likelihood by Category				
Likelihood Category	Weight	Negligible = 1	Unlikely = 2	Possible = 4	Likely = 7	Very Likely = 10
Physical Condition	0.60	Very good (Condition Grade 1)	Good (Condition Grade 2)	Fair (Condition Grade 3)	Poor (Condition Grade 4)	Very poor (Condition Grade 5)
Performance	0.20	Sufficient capacity to meet average and peak flow requirements; appropriate utilization and function	Under-utilized or oversized, causing O&M issues	Sufficient capacity but does not meet functional requirements or over-utilized	Able to meet current average capacity demands but not peak demands	Unable to meet current average capacity needs
O&M Protocols	0.05	Complete, up-to-date, written/online, easily accessible	Complete, written/online, up-to-date, but not easily accessible	Written/online but not complete, not up-to-date, or not easily accessible	Written/online but not complete, out-of-date, or location is unknown	None
Reliability: Planned maintenance as a % of total maintenance	0.15	> 75%	50% to 75%	35% to 50%	25% to 35%	< 25%

The five levels of likelihood listed across the top row of the matrix correlate with the five grades of condition assessment. The levels of likelihood (e.g., negligible, unlikely, possible, likely, and very likely) can be chosen by the Asset Management Team. The score (e.g., 1, 2, 4, 7 and 10) can also be chosen by the Team. However, as with the consequence scoring system, it is suggested that the 1 to 10 scale be maintained.

The cells in the matrix should be completed by the Asset Management Team and be based on the asset management mission and the experience, knowledge, and judgment of the Team. Involvement of the functional teams whose members may be more aware of asset condition and performance may be appropriate during the development of the likelihood matrix.

Assign Consequence Scores to Assets[13]

With the asset hierarchy completed to the fourth or fifth level, and the consequence and likelihood matrices created, assets can be evaluated for consequence of failure. Beginning with the assets at the second level of the hierarchy (e.g., "Treatment Plant 1"in Exhibit 5.7), the consequence matrix should be applied to each of the second-level assets. For example, a second-level asset is chosen and the level of service objectives are attributed to a hypothetical failure of that asset. So, within an Asset Management Team workshop or a workshop of the functional teams, the question is posed, "What is the consequence to the level of service if this asset fails to perform as intended?" The focus of the question is on the first objective, then the second, and so on. The matrix is used to guide the team members in reaching a consensus as to the severity of the consequence due to the failure of that particular asset.

---

[13] Assets also refers to groups of assets that may be combined at higher levels of the hierarchy.

During the scoring process it is assumed that only the asset being assessed fails (i.e., no multiple asset failures). This is where redundancy is considered. Because only the asset under consideration is being scored for consequence, if that asset has redundancy (i.e., another asset can perform its function), then the consequence should be scored considering that the redundant asset will be used to the extent possible to replace the function of the failed asset.

If an asset has failed in the past, it is helpful for someone to discuss the actual causes and consequences of such failure. If the asset has not failed, those staff members familiar with the asset should describe a likely failure mode (an event that could cause the asset to fail) to help the team members score the asset most knowledgeably.

Alternatively to scoring one asset at a time, it may be helpful to score one objective at a time for all second-level assets. Some teams find this easier because they are focusing on one category of consequence at a time. Should a stalemate exist as to which score to apply to an asset, it is suggested that the team choose the higher score. This will result in a higher risk and, therefore, cause a more thorough evaluation of the asset.

Finally, for each asset, the sum product of the weighted scores is calculated to arrive at a consequence score. Exhibit 5.10 is an example of a consequence scoring table.

EXHIBIT 5.10   Consequence Scoring of Assets at the Second Level of the Hierarchy

Asset		Wt	Health & Safety	Compliance with Regulations	Financial Impact	Disruption to the Community	Service Delivery	Ability to Respond and Continue Service	Consequence Score
			0.20	0.20	0.10	0.15	0.20	0.15	
Asset	2-V		10	10	10	10	1	1	6.9
Asset	2-W		7	1	4	1	4	7	4.0
Asset	2-X		7	1	7	1	1	7	3.7
Asset	2 Y		7	4	7	4	7	10	6.4
Asset	2-Z		7	4	7	4	7	7	6.0

Assign Likelihood Scores to Assets

As with consequence, the assets at the second level of the hierarchy should be scored for likelihood of failure. The process is the same. Beginning with the assets at the second level of the hierarchy, the likelihood matrix should be applied to each of the second-level assets. For example, a second-level asset is chosen and the likelihood categories are attributed to that asset. One way to accomplish this is to treat each likelihood category as a question. For example, for each asset ask, "What is the condition of this asset?" and use the matrix to select the answers.

Again, the team may score all the second-level assets against each likelihood category, rather than scoring one asset at a time against all of the categories. Should a stalemate exist as to which score to apply to an asset, it is once again suggested that the team choose the higher score to cause a higher risk and, therefore, a more thorough evaluation of the asset.

For each asset, the sum product of the weighted scores is calculated to arrive at a likelihood score. Exhibit 5.11 is an example of a likelihood scoring table.

EXHIBIT 5.11   Likelihood Scoring of Assets at the Second Level of the Hierarchy

Asset		Wt	Physical Condition	Performance	O&M Protocols	Reliability: Planned Maintenance	Likelihood Score
			0.60	0.20	0.05	0.15	
Asset	2-V		4	2	1	4	3.5
Asset	2-W		2	2	2	1	1.9
Asset	2-X		2	1	1	4	2.1
Asset	2-Y		2	2	2	4	2.3
Asset	2-Z		7	4	2	10	6.6

Calculate Relative Risks of Highest-Level Assets and Prioritize by Risk

Once all of the second-level assets have been scored for consequence and likelihood, the relative risk of each is calculated by multiplying the consequence score and the likelihood score for each asset. The assets can then be ranked by risk score. As described in Chapter 4, it is helpful to group the assets by those that have risk scores within a certain range and assign a risk grade from A through E, with those assets having the lowest relative risks getting an A, and those having the highest relative risks getting a E. Grades B, C and D are assigned to those assets with risk scores in between.

While it is frequently apparent from a quick review of the ranked list of assets how to assign the assets into grades, one method that can be used to mathematically make the assignments is to calculate the difference between an asset and the asset having the next highest risk score. Where the difference is substantial, the risk grade should be changed to the next grade. Exhibit 5.12 shows an example of risk scoring, ranking, and grading.

EXHIBIT 5.12   Risk Scoring, Ranking, and Grading at the Second Level of the Hierarchy

Asset		Consequence Score	Likelihood Score	Risk Score	Risk Rank	Risk Grade
Asset	2-V	6.9	3.5	24.2	2	C
Asset	2-W	4.0	1.9	7.6	5	A
Asset	2-X	3.7	2.1	7.8	4	A
Asset	2-Y	6.4	2.3	14.7	3	B
Asset	2-Z	6.0	6.6	39.6	1	D

Identify Risk Reduction Options

The Asset Management Team and/or utility management must decide how to define unacceptable risk. One method is to designate a risk score above which risk is considered unacceptable and risk mitigation efforts must be pursued. Another method is to consider all assets receiving a risk grade of D or E to be posing an unacceptable risk. The decision is that of the utility.

Once the assets with unacceptable risks are determined, three alternatives are available to the utility:

1. Proceed to the next level of the asset hierarchy, starting with those assets that fall under the second-level asset with the highest risks, and repeat the process at the third-level of the hierarchy.

2. Gather more information, including field condition assessments and performance evaluations, for the high risk assets and re-score the risk based on the advanced information and data.

3. Begin investigating the cause contributing to the high risk scores by reviewing the scoring in the matrices and then identifying and evaluating risk mitigation options.

If alternative 1 is chosen, then all three alternatives are available once the assets at the next level of the asset hierarchy are assessed for risk. When the final level of the initial hierarchy is reached (e.g., fourth or fifth level), alternatives 2 and 3 are available. If alternative 2 is chosen at any point in the risk assessment, alternative 3 becomes the only choice remaining and the process of risk mitigation identification and evaluation begins.

Exhibits 5.13 through 5.15 shows scoring of consequence, likelihood, and risk for third level assets.

EXHIBIT 5.13   Consequence Scoring of Assets at the Third Level of the Hierarchy

Asset		Wt	Health & Safety	Compliance with Regulations	Financial Impact	Disruption to the Community	Service Delivery	Ability to Respond and Continue Service	Consequence Score
			0.20	0.20	0.10	0.15	0.20	0.15	
Asset	3-Z-0100		4	1	7	1	1	2	2.4
Asset	3-Z-0101		4	4	7	7	4	1	4.3
Asset	3-Z-0102		10	10	10	7	4	7	7.9
Asset	3-Z-0103		7	7	10	10	1	4	6.1
Asset	3-Z-0104		4	1	1	1	1	1	1.6
Asset	3-Z-0105		4	1	4	4	4	7	3.9
Asset	3-Z-0106		4	7	4	1	1	2	3.3
Asset	3-Z-0107		4	4	4	1	1	1	2.5
Asset	3-Z-0108		1	1	4	1	1	2	1.5
Asset	3-Z-0109		1	1	1	1	1	1	1.0
Asset	3-Z-0110		4	2	4	2	2	2	2.6

EXHIBIT 5.14   Likelihood Scoring of Assets at the Third Level of the Hierarchy

Asset		Wt	Physical Condition	Performance	O&M Protocols	Reliability: Planned Maintenance	Likelihood Score
			0.60	0.20	0.05	0.15	
Asset	3-Z-0100		1	2	2	10	2.6
Asset	3-Z-0101		1	2	1	1	1.2
Asset	3-Z-0102		4	7	4	2	4.3
Asset	3-Z-0103		4	7	1	1	4.0
Asset	3-Z-0104		1	2	1	1	1.2
Asset	3-Z-0105		7	4	4	10	6.7
Asset	3-Z-0106		7	4	1	7	6.1
Asset	3-Z-0107		2	1	1	4	2.1
Asset	3-Z-0108		2	4	1	1	2.2
Asset	3-Z-0109		2	2	1	2	2.0
Asset	3-Z-0110		2	1	1	1	1.6

EXHIBIT 5.15   Risk Scoring, Ranking, and Grading at the Third Level of the Hierarchy

Asset		Consequence Score	Likelihood Score	Risk Score	Risk Rank	Risk Grade
Asset	3-Z-0100	2.4	2.6	6.2	5	A
Asset	3-Z-0101	4.3	1.2	5.2	7	A
Asset	3-Z-0102	7.9	4.3	34.0	1	D
Asset	3-Z-0103	6.1	4.0	24.4	3	C
Asset	3-Z-0104	1.6	1.2	1.9	11	A
Asset	3-Z-0105	3.9	6.7	26.1	2	C
Asset	3-Z-0106	3.3	6.1	20.1	4	C
Asset	3-Z-0107	2.5	2.1	5.3	6	A
Asset	3-Z-0108	1.5	2.2	3.3	9	A
Asset	3-Z-0109	1.0	2.0	2.0	10	A
Asset	3-Z-0110	2.6	1.6	4.2	8	A

Prioritize Options Based on Risk Reduction and Cost

Risk reduction options should be chosen based on a thorough evaluation of the factors (from the completed consequence and likelihood scorings) that are contributing to the asset risk. Risk reduction options may include one or more of the following:

· Capital rehabilitation

· Capital replacement

· Changes to operation procedures

· Changes to maintenance practices

· Demand management

· Reduction in level(s) of service

· Improvement in response and recovery

...the utility staff must balance the risk reduction-to-cost ratios of options with the total cost of implementing those options, and doing so within the available budget or otherwise planning for additional sources of funding.

Once several possible risk reduction options are identified, their risk reduction-to-cost ratios should be determined. The risk reduction can be calculated by using the consequence and likelihood matrices to score the asset assuming the risk-reduction option was implemented and calculating the resulting risk score. That score, subtracted from the current risk score of the asset quantifies the risk reduction for the particular option. The denominator of the ratio, cost, should be the life-cycle cost of the option.

The risk reduction–to-cost ratios of the identified options provide important guidance in selecting an option for implementation. Generally, the option with the highest risk-reduction-to-cost ratio is preferable and should be the option considered for implementation. However, other considerations may suggest the implementation of an option that does not result in the highest risk reduction-to-cost ratio. Most commonly, this consideration is the need to implement risk reduction options for several high-risk assets within a given budget. Consequently, the utility staff must balance the risk reduction-to-cost ratios of options with the total cost of implementing those options, and doing so within the available budget or otherwise planning for additional sources of funding. Exhibit 5.16 is an example of the way in which a utility might prioritize risk reduction options.

EXHIBIT 5.16   Example of Prioritizing Risk Reduction Options Using the Risk Reduction-to-Cost Ratio

EXAMPLE:	Referring to Exhibits 5.13 through 5.15, assume that the risk grade "D" (with a Risk Score of 34.0) of Asset 3Z-102 is unacceptable, and a risk reduction option must be implemented.
Step 1.	Review consequence and likelihood scoring to identify the categories that are causing the high risk for the asset. From Exhibit 5.13, it appears that this asset has a high consequence of failure, receiving a score of "10" for Health & Safety, Compliance with Regulations, and Financial Impact. It also scored a "7" for Disruption to the Community. From Exhibit 5.14, its likelihood of failure is not as high as some of the other third-level assets, but its Performance scored high (7).
Step 2.	Identify risk reduction options (capital improvements, O&M enhancements, and/or other means) that will reduce this asset's risk of failure. To reduce the consequence of failure, this utility might consider installing a redundant asset. To reduce the likelihood of failure, improving the performance of the asset by more frequent maintenance is also an option. Other options may be identified as well.
Step 3.	Estimate present value (PV) of risk reduction options. Estimate capital and recurring costs over a period of 10 years (estimated useful life of this type of asset) and calculate the present value using appropriate formulas. For this example, let's say Option 1 (installing a redundant asset) has a PV of $500,000, and Option 2 (more frequent maintenance) has a PV of $200,000.
Step 4.	Calculate a "new" risk score assuming options are in place. If Option 1 is in place, assume that the score for two of the consequence categories (Health & Safety and Financial Impact) will be reduced to a 4 and three of the consequence categories (Compliance with Regulations, Disruption to the Community, and Service Delivery) will be reduced to 1, resulting in a total consequence score of 2.8. Because the asset in question has not been changed, the likelihood score stays the same at 4.3 and the new risk score for Option 1 is 12.0.
	If Option 2 is in place, assume that the Performance of the asset goes from a score of 7 to a score of 4, and Reliability is reduced to a score of 1. This results in a likelihood score of 3.6. Because the consequence score with Option 2 remains at 7.9, the new risk score is 28.4.
Step 5.	Calculate the risk reduction-to-cost ratio for each option. Option 1 has a ratio of (34.0-12.0)/$500,000 = .044 units per $1,000. Option 2 has a ratio of (34.0-28.4)/$200,000 = 0.028 units per $1,000. Consequently, even though Option 1 is more expensive, it provides considerably more risk reduction per dollar than Option 2. Similar calculations should be performed for all options identified.

In making the final decisions on the options to implement at any given time, it is useful to once again review all of the completed matrices to see which consequence and likelihood criteria are addressed by each option. All else being equal, utility staff may want choose the options that address the criteria considered to be the most important (i.e., the criteria given the highest weight). Consideration may also be given to implementing the combination of projects that reduce the greatest amount of risk in the aggregate for the budget available.

Perform Multi-Attribute Analysis as Appropriate

The risk reduction-to-cost ratio analysis presented above provides a useful method for selecting capital projects to lower the risk of failure posed by existing assets. However, in preparing a capital improvement plan, driving forces other than current risks that are posed by failure of existing assets must frequently be considered. Drivers such as growth demands, future regulatory requirements, political forces, and joint-infrastructure issues (e.g., pipeline interferences with road construction) cause a utility to identify projects for implementation in addition to those projects identified for asset risk reduction.

Drivers such as growth demands, future regulatory requirements, political forces, and joint-infrastructure issues (e.g., pipeline interferences with road construction) cause a utility to identify projects for implementation in addition to those projects identified for asset risk reduction.

A suggested method of evaluating all identified projects—those for asset risk reduction as well as those to meet other drivers—is multi-attribute analysis (also known as multi-attribute utility analysis, multi-criteria decision analysis, and other like terms). Using multi-attribute analysis, a utility can evaluate numerous drivers for capital projects, including asset risk reduction through a process similar to the consequence/likelihood/risk assessment process described above.

Through multi-attribute analysis, a utility identifies the drivers or attributes that are key to determining the capital projects that are implemented. Such attributes can be risk reduction, risk reduction-to-cost ratio, ability to meet a future regulatory requirement, ability to meet a certain future capacity need, etc. Each of the attributes is given a weight, and by consensus the identified projects are scored against the attributes. The results allow for a ranking of capital projects in the order that they meet the drivers identified by the utility. The AwwaRF *Capital Planning Strategy Manual*[14] presents a detailed methodology for ranking capital projects.

Refine Capital Improvement Plan and O&M Activities

There is no absolute or universal solution for selecting capital projects, enhancing O&M protocols, or refining business processes to assure that asset life-cycle costs are minimized while a utility provides its established levels of service. However, the methodologies described in this chapter provide an effective decision framework to guide utility staffs toward more rigorous, reproducible, and defensible decision making.

---

[14] *Capital Planning Strategy Manual*, American Water Works Association Research Foundation, 2001.

Using these methodologies to supplement the experience, knowledge, and professional judgment of staff members, utilities can more confidently refine their capital improvement plans and implement operations and maintenance activities to optimize the management of their infrastructure assets.

Sustainability through a Continuous Process

Asset management is a continuous process. The methodologies presented must be incorporated into the utility's organization through initial and periodic training, reinforcement by management, and enhancement of other business processes.

The physical condition and performance of assets should be gathered on a regular basis (e.g., as part of preventive maintenance tasks), and the updated information should be used to re-calculate the likelihood of failure and risk on at least an annual basis to guide the next year's planning process and identify needed modifications to O&M protocols.

The methodologies...must be incorporated into the utility's organization through initial and periodic training, reinforcement by management, and enhancement of other business processes....

The entire process, including a review of levels of service and design of the consequence matrix, should be done every five years or sooner if significant changes are made to the utility's strategic plan. Additionally, utilities should take advantage of available technology to help manage the enormous amounts of asset data that will be collected and consider the use of the newer predictive modeling applications to match planning horizons with the long life-cycles of infrastructure assets (see Chapter 6).

CHAPTER 6

Software Applications and System Integration Principles

While smaller utilities may be able to apply the concepts presented in the previous chapters using paper records and common software such as Microsoft® Office, many utilities will want to take advantage of the capabilities of more advanced technology solutions. A wide variety of software applications and hardware are available for utilities to more efficiently store, retrieve and manipulate the large quantities of data associated with water and wastewater infrastructure assets. However, before procuring or implementing such tools it is imperative for utility managers to understand the purpose and limitations of the numerous products being offered, and how they can be most effectively utilized through system integration.

> *...before procuring such (advanced technology) tools it is imperative for utility managers to understand the purpose and limitations of the numerous products being offered...*

Software Applications

Most utility professionals are familiar with some type of asset management software because of its wide use in the operations and maintenance of water and wastewater systems. The most common software applications include:

- Computerized Maintenance Management Systems (CMMS) – to generate work orders and preventive maintenance schedules; track equipment downtime; generate alerts for needed action regarding critical events; track spare parts; maintain asset inspection reports, etc.

- Geographic Information Systems (GIS) to store, display, and use geographically referenced information, that is, data associated with asset locations

- Customer Information Systems (CIS) – to store customer information, generate bills, track customer complaints, maintain meter information, etc.

- Financial Information Systems (FIS) – to provide financial information and reporting about the utility; provide cost information on labor, operating expenses, outside service fees, debt service, project costs, operating and capital budget development, management, and tracking; and maintain current value of utility assets and depreciation for financial reporting

These applications are crucial for the effective operation of utilities, but they are generally retrospective in nature. They store and display information that has been previously input and, in some cases, use that information to generate action items such as work orders, alerts, or reminders. As stand-alone applications, they typically do not have the capability to analyze the information needed to forecast asset conditions, risk, or investment needs.

Over the past several years, significant advances have been made in the development of software that goes beyond the traditional CMMS, GIS, CIS, and FIS applications listed previously. These newer software packages can be used to analyze asset data and provide decision support for operations, maintenance, and capital planning. These packages also incorporate predictive modeling that can be used for the long-term planning and management of assets. However, they are currently not widely used due to the complexity of long-term planning issues, especially the determination of the condition of the assets and, thus, their associated remaining service lives. This is an especially difficult problem for underground assets, and water and force mains in particular, which are not as accessible as aboveground assets or gravity sewers.

These long-term predictive planning software tools primarily use two methods:

· Ranking/point methods, where condition and criticality scores are assigned to prioritize the renewal of the assets

· Statistical survival methods, where the remaining lives of the assets are estimated by analyzing historical survival data (by the use of break, repair, replacement, destructive testing, and nondestructive testing data) and projections are made for their timely renewal

There also are mechanistic models where the condition of the assets is predicted through the use of fundamental engineering equations; however, they are primarily used in academic circles rather than utilities.

In all the methods used, the repair versus renew decisions and the time of renewal are typically determined by the use of economic break-even analysis. Thus, these software tools also include economic analysis tools for completeness.

Several companies in the United States and Canada have developed more advanced asset management software that can be used to forecast asset condition and risk, and allow for the evaluation of capital investment needs based upon condition and risk. There are also a few software products developed by non-profit organizations that also offer predictive modeling capabilities. Exhibit 6.1 provides a summary of some of these predictive tools by non-profit organizations that are applicable to water and wastewater systems.

EXHIBIT 6.1 Summary of Some Asset Management Software by Non-Profit Organizations

Software & Reference	Description
SCRAPS (Sewer Cataloging, Retrieval, and Prioritization System) http://www.werf.org/AM/Template.cfm?Section=Online Tools&Template=/TaggedPage/TaggedPageDisplay.cfm&TPLID=16&ContentID=2611	SCRAPS is a tool for small- to medium-sized wastewater utilities to help assess limited information about sewers and estimate the likelihood and consequence of pipe failure. It uses conditional probabilities to estimate the risk of failure due to structural or operational mechanisms. The tool can be used to strategically focus sewer inspection programs in those areas most likely to need attention.

KANEW (AwwaRF, 1997) http://www.awwarf.org/research/topicsandprojects/execSum/265.aspx	Based on statistical analysis of pipe lifetimes for homogeneous categories of pipes, KANEW can be used to identify appropriate lengths of pipes of different pipe material and sizes to be replaced in each year. As a macro level model, prioritization is not for specific segments of pipe, but for categories of pipe. The AwwaRF version of KANEW runs only on Access 97, while the more up-to-date, now commercial version marketed in Europe can run without database management software.
CARE-W (Computer Aided Rehabilitation of Water Networks) http://care-w.unife.it	CARE-W provides prioritized water main renewal strategies and incorporates hydraulic modeling to assess pipeline reliability in the renewal prioritization methodology. It consists of the following: · A control panel of performance for renewal · Modeling for failure forecasting · Hydraulic service reliability models · A decision support tool for developing an annual renewal program · A long-term strategic planning and investment tool
CARE-S (Computer Aided Rehabilitation of Sewer Networks) http://care-s.unife.it/	CARE-S provides methods and models for the long-term planning, project ranking, and renewal selection for wastewater and stormwater collection systems. It consists of the following: · A performance indicators generator for renewal decisions, including analytical and statistical procedures to forecast performance · A procedure to define the socio-economic and environmental risks of malfunctioning sewer systems · A database for choosing an appropriate rehabilitation technology · A tool for determining the best long-term strategy for renewal · A tool for the assessment of the hydraulic, environmental, and structural conditions of the network including their changes over time · A multi-criteria decision tool for prioritizing renewal projects
PARMS (Pipeline Asset and Risk Management System) - Planning http://www.csiro.au/solutions/psag.html	PARMS-Planning is a suite of computer-based models designed to assist in the management of water supply network assets. Currently the PARMS-Planning model is in use to: · Identify assets for replacement · Assess replacement based upon the predicted number of failures in any one year · Forecast the expected annual number of failures · Calculate the cost implications of different management and operational scenarios to be evaluated.
WARP (Water Mains Renewal Planner) http://irc.nrc-cnrc.gc.ca/ui/bu/warp_e.html	WARP is a software decision support tool for planning of water main renewal. It analyzes the historic breakage rates of water mains, projects future breakage rates, computes life-cycle costs, and generates planning scenarios. It also takes into account time-dependent effects such as climate and cathodic protection.

Halfawy et al. (2006)[15] recommend that asset management software packages should:

· Enable efficient and systematic collection, storage, query, retrieval, management, analysis, and reporting of asset information.

· Integrate and manage various aspects of the asset life-cycle by incorporating different work processes and their associated datasets.

· Enable the sharing of data across the utility and with other parts of the larger organization.

· Increase operational efficiency by aiding in the planning, execution, and coordination of maintenance operations, and tracking and managing the information related to projects, work orders, inspections, etc.

· Assist in coordinating and optimizing the allocation and distribution of maintenance budgets according to the priority and risk associated with deteriorating components of the assets.

Whatever the final decision with respect to asset management software, it is important that the utility remain mindful of two important principles:

1. The needs of the organization should drive the ultimate selection of asset management software and not vice-versa; asset management software is merely a tool.

2. Asset management software cannot ever be perceived as a substitute for prudent management and sound planning.

The implementation of asset management software can be done either enterprise- (utility-) wide or by business units (departments). Under ideal circumstances, the enterprise-wide implementation of asset management software is preferable because the financial aspects of asset management can also be included along with the operational and engineering functions. However, due to the complexity of implementing enterprise-wide software, some utilities prefer to implement it for specific business units. One option is to take an incremental approach where implementation starts with one business unit and expand to others until a more comprehensive (but not necessarily enterprise-wide) system is implemented.

The following advantages have been cited by utility managers for starting the implementation with one business unit and then expanding the program:

· The program is tested in a smaller setting, requiring fewer resources and fewer people to buy in to the system.

· This testing ensures that systems and procedures will work for other business units before the software is implemented there.

· There is increased likelihood of demonstrating early successes, which makes rolling out the software to other business units easier.

· Procedures and tools developed in one business unit can be leveraged at other business units, reducing the level of effort required to implement them at additional business units.

---

[15] Halfawy M R, Newton L A and Vanier D J (2006) "Review of commercial municipal infrastructure asset management systems," ITcon Vol. 11, *Special Issue Decision Support Systems for Infrastructure Management*, pg. 211-224, http://www.itcon. org/cgi-bin/works/Show?2006_16.

Each of the approaches mentioned above have varying cost, resource, and schedule implications and should therefore be carefully evaluated prior to the selection of the asset management software.

There are several categories of features that utility managers need to consider when selecting asset management software. Some of these categories pertain to any type of software, while others are relevant to specifically asset management software.

- Generic categories of features include:

    - Software functionality;

    - Software architecture; and

    - Input and output functionality.

- Asset management-related features are:

    - Asset management components; and

    - Types of assets included.

When evaluating asset management-related features, integration among different asset categories is an important factor that needs to be considered. Lack of integration can create significant inefficiencies in maintenance coordination and asset planning. Ideally, repair, maintenance, or construction activities at common locations should be coordinated to span multiple infrastructure assets as much as possible. This minimizes the disruption, cost, and risks associated with maintenance operations. For example, asset management software that can schedule both water main and sewer repairs provides additional value to utilities.

One of the key considerations in software selection is whether the software should be a commercial off-the-shelf (COTS) package or custom software. COTS applications can be either general purpose asset management software or public works/utility-focused software. Further information on COTS software is presented later in this chapter.

Exhibits 6.2 and 6.3 provide fundamental information about asset management software functionality and system architecture by type. Both exhibits list the advantages and disadvantages associated with each item.

EXHIBIT 6.2   Asset Management Software Functionality

Software Type	Advantages	Disadvantages
COTS general-purpose software, typically procured for financial functions, offers generic functionality to accommodate a wide range of processes, but usually needs to be customized for specific applications through add-on modules.	Single software vendor for all functions throughout the organization. Lower costs for utility to purchase modules if the parent organization already has the general-purpose software.	Significant installation and startup costs. COTS modules may not fully accommodate the specific needs of the utility. Relatively high maintenance costs.
COTS public works/utility-focused software provides a set of built-in data models and processes to support the management of specific classes of public works assets (e.g., treatment plants, sewers, roads, etc.).	Already customized for needs of the utility. Typically based on an asset hierarchy with manipulation of data possible by asset category and individual assets.	May not meet highly specialized needs.
Custom software is an alternative if the functionality offered by COTS asset management software is either too limited or too extensive, making it complicated to use.	Customization enables utilities to use legacy software and assure their specific needs are met.	Customization can be complex and expensive. Integration with other software acquired in the future can be problematic.

EXHIBIT 6.3   Software Architecture

Architecture Type	Advantages	Disadvantages
Client-server is a form of network architecture where the user's computer is considered a client, and the asset management databases and applications that the client accesses on a LAN or WAN are considered the server. When the user queries information on a certain asset, the server finds the information and sends it back to the user.	Databases are easier to maintain and more cost effective in multi-user environments.	More vulnerable to security breaches.
Desk-top based architecture requires that the software and all data be installed on each computer where the application will be used.	More secure because of limited access.	Maintenance of up-to-date databases is more onerous. A common data set is not instantly accessible to all users or available for update by all users. Furthermore, it has limited and specific functionality with regard to other software integration.

Exhibit 6.4 lists the input and output features that need to be considered when selecting asset management software. Typically, the more features, the more functional the software. However, more functionality is usually at the risk of increased complexity.

EXHIBIT 6.4   Data Input and Output Functionality

Category	Description
Data Import and Export Capabilities	The ease with which data and various types of files can be imported and used. Similarly, the ease with which output files can be exported and used by other applications.
Reporting Capabilities	The capability of having both a range of pre-formatted reports and the option of generating custom reports. Pre-formatted reports provide users with a consistent set of forms for different assets, enabling users to become familiar with the software relatively quickly. Similarly, the ease of generating custom reports provides further flexibility.
Integration with GIS and CMMS	Some software packages use a GIS interface and GIS database to maintain and integrate asset data. Others interface with a CMMS database. This integration feature is critical to ensuring consistency of asset data and eliminating multiple databases that must be updated separately. On the other hand, some software packages maintain two databases and spatially link work orders and service requests to specific assets or to street addresses. This approach enables utilities with an incomplete asset inventory to link work orders and service requests to addresses.
Interface with External Applications	The capability to interface with external applications such as GIS, email, or Supervisory Control and Automatic Data Acquisition (SCADA) systems. An important related capability is the flexibility to interface with legacy software used by the utility. Otherwise, abandonment or conversion of the legacy software can be costly.
Remote Access	The capability to support field operations by enabling browser-based, wireless Internet access to the asset database and allowing field staff to access and update work orders and service requests and view infrastructure maps. Access can be through laptop computers or PDAs.

Asset management software systems have various modules that address some, if not all, of these functional requirements. According to Halfawy et al. the majority of the existing systems focus primarily on supporting day-to-day activities, and only a small number support long-term renewal planning. Also, many fundamental asset management functions, such as performance modeling and maintenance prioritization, are not supported by most of these applications.

The *International Infrastructure Asset Management Manual* (2006) lists the typical modules that are provided in basic and advanced asset management software systems (Exhibit 6.5). A description of these modules is briefly outlined in Exhibit 6.6.

EXHIBIT 6.5   Basic and Advanced Asset Management Software Modules

Basic Modules	Advanced Modules
Asset inventory, accounting, and control Maintenance management Contract management Job/resource management Condition assessment	Predictive modeling Risk assessment Life-cycle costing Alternative evaluation Project planning and optimization

EXHIBIT 6.6   Asset Management Software Modules

Category	Description
Inventory	The inventory modules serve as the asset database and provide the functionality to search and retrieve asset data. Some modules have a hierarchical structure that allows the user to specify the level of data detail required for each asset. This enables the user to drill down through the asset from facility level to component and element levels. For software systems that have GIS capability, the asset inventory can be viewed both geographically and by asset category. Asset inventory modules are typically tied to condition assessment modules, which also include inspection information.
Condition Assessment	Condition assessment modules include asset inspection information and allow users to select an existing assessment protocol or to define their own protocols. Some modules utilize weighting factors (or ranking scores) and assign weights to a variety of assessment metrics related to their condition and criticality. Condition assessment modules are typically integrated with analysis and evaluation tools to predict the future condition of the assets and allow development of renewal plans as necessary. Some software packages can import condition assessment data from other sources and directly link these data to assets in the inventory.
Valuation	Some modules have a built-in valuation capability that uses economic factors to calculate asset values. In some cases, these factors can be amended by the user.
O&M Management	These modules typically link work orders to assets in the inventory, set up recurring schedules for preventive maintenance activities, email notices to any email program, view schedules, generate reports by maintenance function or assets, track project costs, and support activity-based costing.
Analysis and Evaluation	There are numerous analysis and evaluation tools that can be used for decision support. Condition assessment and asset survival rate data can be used to predict the remaining useful lives of assets and, thus, help develop renewal plans. Some software systems have modules that can be used for asset valuation, determination of deferred maintenance, condition assessment, estimated remaining service life, and prioritization of maintenance and rehabilitation processes.

System Integration Principles

Within the context of asset management, key data management and integration principles need to be established and implemented to obtain the maximum benefit from the investments in information technology. Exhibit 6.7 lists ten guiding principles (followed by an overview of each) that support effective application integration and data management. When these principles are followed, there is a high probability that a utility can efficiently implement and maintain effective asset management.

EXHIBIT 6.7   Guiding Principles for Effective Application Integration and Data Management

1. Information technology (IT) systems are assets that should be managed like other assets.

2. A robust enterprise architecture yields significant benefits.

3. Enterprise business processes should be linked to key strategic outcomes.

4. Business applications should be rationalized where possible.

5. Commercial off-the-shelf (COTS) applications should be utilized when and where possible.

6. An asset register facilitates asset identification.

7. Business improvement methodologies should be incorporated.

8. A well-designed performance measurement system allows for continuous improvement.

9. Information and data views are most effective when aligned with user needs.

10. Effective data security improves asset performance and protects the utility and customers.

1. IT systems are assets that should be managed like other assets. Many utilities have made significant investments in information technology (IT) systems. IT has been seen as one important method of improving the effectiveness and efficiency of utility operations and obtaining productivity gains. Because of this substantial investment, IT systems are an important asset and should be managed like any other major asset. When this philosophy is in place and practiced, the goals and objectives envisioned in the original business case for implementing these IT solutions are more likely to be realized. Exhibit 6.8 demonstrates the concept of life-cycle approach to managing IT investments.

EXHIBIT 6.8   Total Life-Cycle Approach for IT Assets

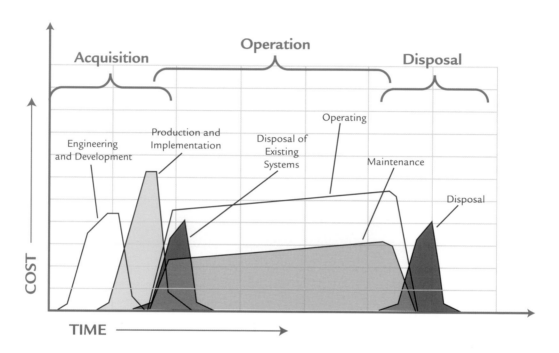

The typical process of identifying needs, investigating options, and procuring a technology solution is commonly used in utilities. The upfront costs of introducing business solutions into an organization can be substantial. However, many utilities forget to manage and account for ongoing O&M as well as the disposition costs, which can be even greater than the upfront procurement and implementation costs. Taking a life-cycle cost perspective enables utility managers to obtain the intended business results and keep costs of IT solutions from spiraling out of control.

2. A robust enterprise architecture yields significant benefits. Since the creation of the first two separate business applications, information sharing has been an ongoing need and a challenge. Responding to this challenge is how best to achieve integration in a speedy, cost-effective, and flexible manner. In addition, it has become increasingly important to minimize total life-cycle costs for developing and maintaining integrated solutions. Enterprise architecture can be defined as having a common approach and standards for business applications, data, and networking/computing infrastructure. When enterprise architecture is in place, utilities will find they can more easily adapt to change, incorporate new applications faster into the business, improve their overall service level performance to customers, and reduce life-cycle IT costs. Exhibit 6.9 demonstrates the concepts of enterprise architecture and its key attributes.

EXHIBIT 6.9   Concepts of Enterprise Architecture

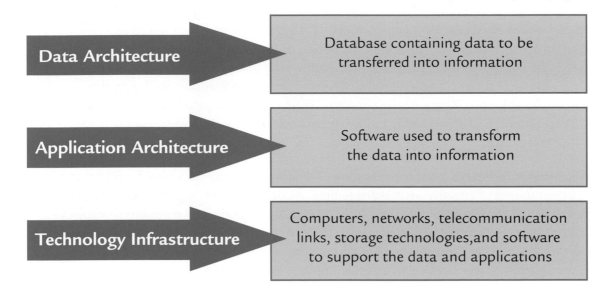

Data Architecture ➤	Database containing data to be transferred into information
Application Architecture ➤	Software used to transform the data into information
Technology Infrastructure ➤	Computers, networks, telecommunication links, storage technologies, and software to support the data and applications

3. Enterprise business processes should be linked to key strategic outcomes. Well-known companies like eBay, Yahoo, Expedia, Dell Computers, and other world class, service-oriented companies have created enterprise business processes that their customers and employees utilize without having to know about all of the technology being deployed in the background. These enterprise workflows provide the customer/end user with the right level of information in a timely manner to perform an activity or work function. Utilities can take these same well-established concepts and apply them to their organizations to achieve maximum benefits for their customers and employees.

4. Business applications should be rationalized where possible. It is likely that multiple business applications exist that perform the same or similar functions. It is not uncommon to find two or more CMMS packages being used by a utility or multiple SCADA and GIS systems in place. This duplication of applications and asset data sets can result in much higher life-cycle costs for IT solutions, unreliability of data, and less than desirable results for customers and other stakeholders. A key focus should be on rationalizing business applications where possible and eliminating duplication of applications and data sets. Utilities that rationalize business applications are more likely to have better information sharing between departments, recognizable improvements in overall data management, and enhanced service level results for customers.

5. Commercial off-the-shelf (COTS) applications should be utilized when and where possible. Over the past decade, COTS applications for asset management, operations and maintenance activities, laboratory information management, and customer billing have become mainstays in the marketplace. Although there may be some exceptions that require building customized applications, COTS applications are seen as less risky alternatives. Although in some cases, a utility must modify its business practices to accommodate the use of COTS, most applications are based upon industry best practices and thus provide utilities the impetus to make these business process changes. In addition, most of the major COTS applications provide some level of integration with other COTS applications (e.g., CMMS to GIS).

An alternative approach of utilizing "hosted solutions" (more commonly referred to as application outsourcing) provides utilities with an additional option of licensing business applications from a third party. The benefit of this approach is that it allows the utility to focus on its core business of providing services to internal and external customers rather than focusing on day-to-day management of IT solutions. This evolving option must be evaluated by each utility, but it does provide a potential method of improving customer service levels as well as reducing the life-cycle costs of IT investments.

6. An asset register facilitates asset identification. A uniform protocol for asset identification is a mandatory component of effective asset management. As described in Chapter 3, asset information should be contained in an asset register that is maintained as a database. To assure consistency of asset information and efficiency of maintaining the asset data, it is preferred to have a single database from which all other software applications draw information. However, if this is not practical, the asset database can be part of other applications such as a computerized maintenance management system (CMMS) for vertical assets (e.g., structures, equipment, etc.) and a geographic information system (GIS) for linear assets (e.g., pipelines). Under such an alternative configuration, all applications should synchronize with the primary asset database to assure all data used are consistent and current. Each asset should be identified by a unique asset number or identifier code that should be linked to the parent asset and possibly other descriptors such as geography. Overall, asset identification systems should be as logical and simple as possible.

7. Business improvement methodologies should be incorporated. Effective integration of applications and management of data is not just dependent upon hardware and software. It is critical that a utility's business processes be enhanced and aligned to ensure the most efficient operations and responsive services for its customers, employees, and other stakeholders. Utilities seeking to improve their business operations can utilize a variety of methods and techniques that have helped other public and private organizations improve the results for their customers, stakeholders, and communities. Organizations that attempt to change business operations without a technique or method typically struggle as both theory and practice are necessary to achieve desired results. Several excellent references are available and include the following from the National Association of Clean Water Agencies (NACWA) and the Association of Metropolitan Water Agencies (AMWA):

- *Thinking, Getting and Staying Competitive*

- *Creating High Performance Business Services*

- *The Changing Workforce ... Crisis & Opportunity*

- *The Changing Workforce ... Seizing the Opportunity*

8. A well-designed performance measurement system allows for continuous improvement. In addition to employing IT solutions to better manage infrastructure assets, technology can also be used to facilitate performance measurement and reporting at all levels in an organization. Well-designed, properly implemented performance measurement systems can enable utilities to achieve higher levels of efficiency, quality, and effectiveness. Utilities should select key performance indicators (KPIs) that will permit evaluation of business and technical operations, proper regulatory reporting, and adequate information for their customers and other stakeholders. KPIs should be chosen so that their measurements can be "rolled up" to allow management to compare actual service levels with the target levels of service for the utility. More information on performance measurement can be found in the references listed under "Levels of Service and Performance Assessment" in Appendix B.

9. Information and data views are most effective when aligned with user needs. One of the key challenges faced by all utilities is defining the information and data that are important. Although intentions may be good in capturing as much data as possible, the result from this data overload philosophy is that all information is captured and stored away in databases; very little of it is used, it is difficult to locate, and the overall value of the data diminishes and the cost of maintaining data and information sky rockets. This is the same reasoning for implementing an asset management strategy that includes the Top-Down Approach discussed in Chapter 5.

As depicted in Exhibit 6.10, the determining factor of whether the information and data are important is a matter of how they will be used in the organization, as well as a function of time.

EXHIBIT 6.10   Information and Data Requirements by Role in the Organization and Time

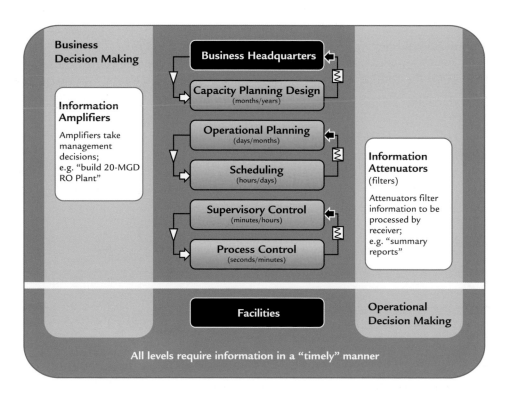

For example, asset performance data are initially created at a facility level and roll up to the business headquarters, which is less focused on operational decision making and more attuned to overall business decision making. Employees in different positions throughout the utility require information at different intervals. For example, capital planners making decisions about the need for a new treatment facility require a longer timeframe to make well-informed capital planning and prioritization decisions. On the other hand, operations—which makes decisions in increments of seconds, minutes, and days—also needs information in a timely manner, but in more discrete summary reports or event notifications. The common need in all levels within a utility is that information is required in a timely manner.

> *One of the key challenges faced by all utilities is defining the information and data that are important.*

10. Effective data security improves asset performance and protects the utility and customers.
Being responsible for a major portion of the nation's critical infrastructure, utilities must protect data and information from unauthorized access and cyber attack to ensure a continuous safe supply of drinking water for its customers and prevent adverse public health and environmental impacts. Utilities are also obligated to maintain a secure database of customer information, as well as protect their financial well-being. As stated in the *Interim Voluntary Security Guidance for Wastewater/Stormwater Utilities*[16], "Information system failure can have catastrophic repercussions to a utility. Compromise of the financial system can result in millions of dollars of lost revenue. Corruption or destruction of operational data can lead to fines due to late or inaccurate regulatory reporting. A sabotaged Web site has the potential to shake public trust during a time of crisis. Interruption of plant processes because of a SCADA system malfunction can lead to a wide range of health implications for the community."

To mitigate the threats of hacking and cyber attacks, utilities must focus on training staff, implementing and enforcing security policies, and installing software and hardware security defenses. Both the *Interim Voluntary Security Guidance for Wastewater/Stormwater Utilities* and the *Interim Voluntary Security Guidance for Water Utilities*[17] have chapters dedicated to cyber security.

---

[16] *Interim Voluntary Security Guidance for Wastewater/Stormwater Utilities.* American Society of Civil Engineers, American Water Works Association, Water Environment Federation, 2004.

[17] *Interim Voluntary Security Guidance for Water Utilities.* American Society of Civil Engineers, American Water Works Association, Water Environment Federation, 2004.

APPENDIX A

Forms and Templates

Asset Management Team

Representative From	Name	Title	Phone
Director or deputy director's office			
Utility system management			
Water			
Wastewater			
Treatment plant supervisor(s)			
Water			
Wastewater			
Maintenance supervisor(s)			
Water			
Wastewater			
Field supervisors			
Distribution system			
Collection system			
Engineering management			
Capital planning lead			
Financial manager or analyst			
IT manager or analyst			
Customer services manager			
Public information officer			
Regulatory compliance officer			
Other			

NOTE: One Team member should be designated as the Asset Management Coordinator and serve as the Team's leader.

Asset Management Mission and Goals

Organizational Mission Statement

Organizational Vision Statement

Asset Management Mission Statement

Asset Management Goals

Levels of Service

Water System Levels of Service	Target Level

Utility-Wide Levels of Service	Target Level

Utility-Wide Levels of Service	Target Level

Consequence Levels by Category

Consequence Levels by Category	Category	Norm Wt	Negligible = 1	Low = 4	Moderate = 7	Severe = 10

Likelihood Levels by Category

Likelihood Levels by Category						
Category	Norm Wt	Negligible = 1	Unlikely = 2	Possible = 4	Likely = 7	Very Likely = 10

Guidelines for Consequence Categories

1. Consequence categories should be based on the organization's established levels of service.

2. Consequence categories should be unique, that is, each category should address criteria not addressed in any of the other categories.

3. When scoring assets for consequence of failure, consider only the criteria associated with the category for which the asset is being scored. Do not have criteria from other categories affect the scoring.

4. When scoring assets for consequence of failure, consider the asset failing in its purpose of performance, not failing due to a catastrophe such as a complete structural collapse or explosion.

5. When scoring assets, consider only a failure of the asset being assessed, not concurrent failure of other assets.

6. Examples of consequence categories include the following (some may be combined or not used as determined by the Asset Management Team):

   · Health and safety of the public

   · Health and safety of employees

   · Environmental impact

   · Financial impact

   · Regulatory compliance

   · Public image

   · Community impact

   · Time to restore service

   · Specific technical service delivery issues

Guidelines for Likelihood Categories

1. Likelihood categories should be unique, that is, each category should address criteria not addressed in any of the other categories.

2. When scoring assets for likelihood of failure, consider only the criteria associated with the category for which the asset is being scored. Do not have criteria from other categories affect the scoring.

3. Examples of likelihood categories include the following (some may be combined or not used as determined by the Asset Management Team):

   · Physical condition

   · Capacity

   · Utilization

   · Functionality

   · Maintenance history

   · Effectiveness of O&M protocols

   · Availability of spare parts

   · Skill level of staf

Risk Reduction Evaluation Form

Asset Posing Unacceptable Risk

Asset ID	Asset Common Name/Location	Risk Score

Primary Cause of Risk

Type	Category
☐ Consequence ☐ Likelihood	

Secondary Cause of Risk

Type	Category
☐ Consequence ☐ Likelihood	

Risk Reduction Option #1

Addresses	Description	Risk Score w/ Option	Reduction in Risk Score	Estimated Lifecycle Cost	Risk Reduction Cost
☐ Primary ☐ Secondary					

Risk Reduction Option #2

Addresses	Description	Risk Score w/ Option	Reduction in Risk Score	Estimated Lifecycle Cost	Risk Reduction Cost
☐ Primary ☐ Secondary					

Risk Reduction Option #3

Addresses	Description	Risk Score w/ Option	Reduction in Risk Score	Estimated Lifecycle Cost	Risk Reduction Cost
☐ Primary ☐ Secondary					

Notes

APPENDIX B

References and Additional Information

Further information about asset management for water and wastewater utilities can be found in several publications and online resources. Additionally, research organizations such as the Water Environment Research Federation (WERF), the American Water Works Association Research Foundation (AwwaRF), the National Research Council Canada (NRC), and the Australian Commonwealth Scientific and Industrial Research Organization (CSIRO) have ongoing projects that investigate new approaches and technologies for improving infrastructure assessment and asset management.

The following is a selection of asset management references and some websites (current as of the publication date of this document). References are listed by category and interested readers can use them for further information. Furthermore, they provide short descriptions of their contents. Relevant ongoing projects that have not yet been published are also listed.

Comprehensive Asset Management

Managing Public Infrastructure Assets to Minimize Cost and Maximize Performance **(The Asset Management Handbook), NACWA, 2002.** Focuses on a strategic approach to asset management, specifically for water and wastewater utilities.

International Infrastructure Management Manual, **Association of Local Government Engineering New Zealand and the Institute of Public Works Engineering of Australia, Version 3.0, 2006.** Addresses asset management concepts and practices for numerous types of infrastructure assets and includes several brief case studies.

Water Infrastructure: Comprehensive Asset Management Has Potential to Help Utilities Better Identify Needs and Plan Future Investments, **GAO, 2004.** Examines (1) the potential benefits of comprehensive asset management for drinking water and wastewater utilities and the challenges that could hinder its implementation and (2) the role that the federal government might play in encouraging utilities to implement asset management.

Water Infrastructure at a Turning Point: The Road to Sustainable Asset Management, **AWWA, 2006.** Provides foundational knowledge of the issues surrounding aging water infrastructure and offers everyday analogies to managing water infrastructure assets.

A Primer on Municipal Infrastructure Asset Management, **NRC, 2004.** Covers some of the fundamentals of asset management for various types of civil infrastructure.

Reinvesting in Drinking Water Infrastructure: Dawn of the Replacement Era, **AWWA, 2001.** Based on a national survey of water utilities, describes the challenges utilities are facing due to aging buried infrastructure. Describes the steps to take to address the problems.

***Asset Management Planning and Reporting Options for Water Utilities,* AwwaRF, 2006.** Provides an understanding of the benefits and drawbacks of various asset management strategies, from basic to high end. Compares case studies in asset management reporting.

***Assessing the Future: Water Utility Infrastructure Management,* AWWA, 2002.** Provides guidance in assessing what and where infrastructure will break down, how and when the breakdown will be addressed, and who will pay for it. Provides guidance on many aspects of water utility infrastructure asset management.

***Capital Planning Strategy Manual,* AwwaRF, 2001.** Provides methods for utilities to improve the processes and tools used to identify, prioritize, and implement capital projects.

***SIMPLE - Sustainable Infrastructure Management Program Learning Environment,* WERF, 2006.** Web-based tool assists wastewater utilities in developing a life-cycle asset management system. http://www.werf.org/AM/Template.cfm?Section=Online_Tools&Template=/TaggedPage/ TaggedPageDisplay.cfm&TPLID=16&ContentID=2611

***Research Priorities for Successful Asset Management: A Workshop,* WERF, 2002.** Reports on a workshop convened in 2002 to characterize the spectrum of asset management programs across water and wastewater utilities and recommend a research agenda to promote the next generation of asset management methods and tools.

***Decision Support Systems for Wastewater Facilities Management,* WERF, 2004.** Describes decision support concepts and discusses the ways in which this methodology could be applied to improve the management of wastewater facilities.

***Development of a Strategic Planning Process,* AwwaRF, 2003.** Presents guidance for strategic business planning that integrates ongoing utility planning activities including asset management, capital improvement planning, integrated resource planning, competitiveness enhancement, and revenue enhancement.

***Distribution Infrastructure Management: Answers to Common Questions,* AwwaRF, 2001.** Provides a practical overview of research on infrastructure management and describes the elements of a comprehensive infrastructure management program, pointing to references for solving various infrastructure problems.

***Strategic Planning and Organizational Development for Water Utilities,* AwwaRF, 2004.** Presents an overall strategic planning framework, incorporating various corporate-level strategies and business tools that can be customized by individual utilities.

***Geographic Information Systems (GIS) and Interoperability of Software for Municipal Infrastructure Applications,* NRC, 2005.** Provides a detailed description of GIS as an asset management tool, and presents a comprehensive evaluation of the state-of-practice of GIS in municipalities.

***Journal of Infrastructure Systems,* ASCE.** Publishes cross-disciplinary papers about methodologies for monitoring, evaluating, expanding, repairing, replacing, financing, and sustaining the civil infrastructure.

GIS Implementation for Water and Wastewater Treatment Facilities, **WEF (published by McGraw-Hill), 2004.** Describes the applicability of GIS to water and wastewater systems and incorporating GIS to all aspects of plant operations, maintenance, and management, including its use for asset management.

Managing the Water and Wastewater Utility, **WEF (Dolan, R.J.; Rose, T. D.; Baker, R. A.; Barnes, M. J.), 2003.** A book that includes information about leading and managing in a changing environment, including the incorporation of asset management.

Wastewater Collection Systems Management, **WEF, 1999.** A Manual of Practice that includes an extensive chapter on asset management.

[Online Journal] *Water Asset Management International,* **IWA.** An international newsletter focusing on asset management in water and wastewater utilities. http://www.iwaponline.com/wami/toc.htm

[Online Resource] The **Asset Management Source for North America (AMSNA)** provides information on conferences and materials associated with infrastructure asset management. www.amsna.org (accessed February 5, 2007)

[Online Journal] *AMSNA Review,* **AMSNA.** Publishes news, practices, research, and market information on asset management. http://www.amsna.org/AMSNAReview.html

[Online Resource] **InfraGuide** is a Canadian network of experts and a collection of best practice publications for core infrastructure. www.infraguide.ca

[Online Resource] **Municipal Infrastructure Investment Planning (MIIP).** A website sponsored by the National Research Council (NRC) Canada with a stated objective "To identify, evaluate and develop tools, procedures, and practices that will help infrastructure managers make strategic and cost-effective planning and management decisions." http://irc.nrc-cnrc.gc.ca/ui/bu/miip_e.html

[Online Resource] **USEPA Office of Water, Asset Management Website.** http://www.epa.gov/owm/assetmanage/

[Online Resource] **Water Infrastructure Network (WIN)** is a "broad-based advocacy coalition of local elected officials, drinking water and wastewater service providers, state environmental and health administrators, engineers, and environmentalists for America's drinking water and wastewater infrastructure." www.win-water.org.

[Online Resource] *Asset Management Quarterly, International* **(AMQI).** www.amqi.com

Strategic Asset Management (SAM), **AMQI.** Publishes articles on various strategic asset management issues. http://www.amqi.com/frame.htm

Levels of Service and Performance Assessment

Selection and Definition of Performance Indicators for Water and Wastewater Utilities, **AwwaRF, 2003.** Discusses performance indicators for utilities related to (1) organizational development, (2) customer relations, (3) business operations, (4) water operations, (5) wastewater operations.

Performance and Whole Life Costs of BMPs and SUDS, **WERF, 2005.** Covers best management practices (BMPs) and sustainable urban drainage systems (SUDs), specifically on retention ponds, extended detention basins, vegetated swales, bioretention, porous pavements, and various infiltration practices.

Performance Indicators for Water Supply Services, **IWA. 2006.** Provides guidelines for implementing appropriate performance indicators for various water utility assets. Also provides implementation procedures and case studies on how to adapt the IWA concepts and indicators to specific contexts and objectives.

Performance Indicators for Wastewater Services, **IWA, 2003.** Provides guidelines for implementing appropriate performance indicators for various wastewater utility assets. Also provides implementation procedures and case studies on how to adapt the IWA concepts and indicators to specific contexts and objectives.

Developing and Implementing a Performance Measurement System: Volumes I and II, **WERF, 2000 and 2005.** Presents methods and tools to develop and implement a performance measurement system. Volume I presents secondary research identifying best practices from other industries and how to apply them for water and wastewater utilities. Volume II describes three pilot projects demonstrating the application of Volume I concepts.

Condition Assessment, Failure and Risk

Development of a Tool to Prioritize Sewer Inspections (SCRAPS), **WERF, 2003.** Describes a tool for identifying pipelines at risk of structural and operational failure. Called SCRAPS (Sewer Cataloging, Retrieval, and Prioritization System), the tool lets small- to medium-sized wastewater utilities estimate the probability and consequence of pipe failure.

An Examination of Innovative Methods Used in the Inspection of Wastewater Systems, **WERF, 2004.** Describes common defects found in sewers and methods used to investigate the condition of sewers.

Water Treatment Plant Infrastructure Assessment Manager, **AwwaRF, 2002.** Presents a program and manual for use by utilities in developing strategic plans for water treatment plant infrastructure maintenance and renewal, including condition assessment procedures.

Techniques for Monitoring Structural Behavior of Pipeline Systems, **AwwaRF, 2004.** Reports on techniques that could be used to continuously monitor the structural performance of water distribution systems, primarily large-diameter mains.

Risk Management of Large-Diameter Water Transmission Mains, **AwwaRF, 2005.** Provides utilities with a tool to optimize the frequency of inspection and condition assessment of large water mains to minimize the risk of main failure.

An Evaluation of Condition Assessment Protocols for Sewer Management, **NRC, 2005.** Reviews a condition assessment protocol developed by the Water Research Centre (WRc) and compares this protocol with guidelines developed by the NRC. A detailed comparison of defect coding systems and condition grading systems for the different protocols are discussed. The report also compares maintenance priority rankings based on criticality and consequence of failure.

Software to Prioritize Wastewater Asset Failure and Security Risks, **WERF, 2005.** Addresses two competing needs of wastewater utilities: capital reinvestment for utility assets and investment in security. The software incorporates asset management capabilities into VSAT™, the utility vulnerability self-assessment tool.

Customer Acceptance of Water Main Structural Reliability, **AwwaRF, 2005.** Reports on a methodology for utilities to assess customer perceptions, attitudes, and expectations, and presents a format for quantifying these factors for incorporation into an infrastructure decision-making process.

Nondestructive, Noninvasive Assessment of Underground Pipelines, **AwwaRF, 2002.** Evaluates the feasibility of developing technologies that can be used by water utilities to assess the condition of buried pipelines without excavation and coupon sampling.

Workshop on Condition Assessment Inspection Devices for Water Transmission Mains, **AwwaRF, 2004.** Presents a literature review of non-interruptive condition assessment devices for large diameter transmission mains and reports on the results of a workshop that focused on buried-asset condition assessment research needs and performance of existing technologies.

Operations and Maintenance Management

Tools for Improving Wastewater Operations, **WERF, 1999.** Provides a practical "how-to" guide for improving plant performance and reducing service costs. It features a searchable database of traditional and innovative practices, and includes case studies.

Benchmarking Wastewater Operations: Collection, Treatment, and Biosolids Management, **WERF, 1997.** Provides guidance to enhance performance and efficiency through benchmarking, the systematic process of searching for best practices, innovative ideas, and effective operating procedures. Includes case studies.

Effective Practices for Sanitary Sewer and Collection System Operations and Maintenance, **WERF, 2003.** Online toolkit and CD-ROM to improve operations and maintenance practices, management, and capacity assurance of collection systems.

Decision Support Systems for Wastewater Facilities Management, **WERF, 2005.** This project identifies practices, tools, and methodologies for deriving greater value from the data at WWTPs and incorporating decision support systems into facilities' operational and decision-making processes.

Applicability of Reliability-Centered Maintenance in the Water Industry, **AwwaRF, 2006.** Assesses how utilities could apply RCM to new and existing infrastructure and presents an evaluation of the costs and benefits of such a program.

Materials and Repair, Rehabilitation and Replacement Methods

***Methods for Cost-Effective Rehabilitation of Private Lateral Sewers,* WERF, 2006.** Describes problems related to failing laterals and available options for their inspection, evaluation, and repairs. Also addresses the associated financing and legal issues.

***New Pipes for Old: A Study of Recent Advances in Sewer Pipe Materials and Technology,* WERF, 2000.** Identifies and evaluates new and alternative materials and techniques used in sewer systems and considers the design implications that have arisen.

***Criteria for Renovation or Replacement of Water Treatment Facilities,* AwwaRF, 1991.** Provides guidance and methodology to assist water utilities in comparing construction of a new plant to the rehabilitation of an existing plant, taking into consideration costs, water quality, versatility, and flexibility.

***External Corrosion and Corrosion Control of Buried Water Mains,* AwwaRF, 2005.** Identifies various causes of external corrosion, and presents methods that aid in the identification process. Also determines economical solutions for each type of corrosion and reports on their verification through field trials.

***Assessment and Renewal of Water Distribution Systems,* AwwaRF, 2004.** Synthesizes the knowledge base on condition assessment, repair and rehabilitation, and prioritization. Covers a broad range— from technology-based assessment tools for operators to capital strategy planning methods.

Ongoing Projects

***Develop Protocols for Assessing the Condition and Performance of Water and Wastewater Assets,* WERF and AwwaRF.** This study focuses on measures, metrics, and protocols for assessing asset condition and performance in water and wastewater utilities.

***Asset Management Strategic Planning and Implementation Guidelines for Wastewater Infrastructure,* WERF.** Focuses on effective communications of the overall asset management strategy, both internally and externally.

***Asset Management Research Needs Roadmap,* AwwaRF.** The purpose of the Research Needs Roadmap project is to help AwwaRF define priorities for funding research in asset management. The project includes a white paper that reviews the status of asset management and identifies research gaps, a workshop where participants provide input on future research, case studies, and a final report.

APPENDIX C

Glossary

Asset

An item with an independent physical and functional identity that has a commercial or exchange value and will provide the owner with some form of future benefit. Asset management for water and wastewater utilities generally focuses on capital or infrastructure assets.

Asset group

An aggregation of similar and/or related assets that have similar characteristics such as function, material, size, age, and/or condition.

Asset hierarchy

A representation of the relationships between infrastructure assets. It is arranged as a family-tree in a parent-child format.

Asset inventory

A list of assets that includes details on the type, size, materials of construction and other attributes of each asset, or the process of developing a list of assets including details on type, size, materials of construction, etc.

Asset management

An integrated set of processes to minimize the life-cycle costs of infrastructure assets, at an acceptable level of risk, while continuously delivering established levels of service.

Asset management unit

See "Asset group."

Asset performance

An asset's ability to meet its intended purpose.

Asset register

A written or electronic database containing an asset inventory.

Bottom-up approach

A data intensive approach to asset management based on collecting characteristics of all the utility's assets (e.g., type, material, manufacturer, size, capacity, etc.), a review of operational and maintenance records, field condition assessments, performance analysis, estimation of remaining useful life, determination of asset replacement costs, and other detailed information. All the asset attribute information is then entered into a database to create an asset register.

CADD (computer aided design and drafting)

A software tool that assists users to create information in graphical and digital forms for design and drafting activities and analysis.

Capital asset

A long-term asset that is used to conduct the utility's business; typically includes fixed assets such as land, buildings, other structures, pipelines, machinery, equipment, and fixtures.

Capital renewal

A term encompassing rehabilitating or replacing capital assets to restore an asset, facility, or system to a level equivalent to its original condition and function.

CIS/CRM (customer information system/ customer relationship management)

Computer applications allowing organizations to manage customer information and relationships and predict customer behavior based on databases that store prior interactions with those customers.

Client-server architecture

A form of computer network design where the user's computer is considered a client, and the asset management databases and applications that the client accesses on a LAN or WAN are considered the server.

CMMS (computerized maintenance management system)

A software tool that provides a systematic approach to maintenance planning, budgeting, and work that assists in organizing and analyzing data on asset inventory, condition, and maintenance activities.

COTS (commercial-off-the-shelf) software

Software products that are ready-made and available for sale to the public.

Condition assessment

Continuous or periodic assessments, either judgmental or definitive, to evaluate the physical state of an asset. The results of a condition assessment should be a replicable determination of the objective status of an asset or group of assets.

Condition grade or index

An indictor of the physical condition of an infrastructure asset condition.

Condition-based maintenance

See "Predictive maintenance."

Consequence

The impact on the level of service, utility, customers or general public resulting from an asset failure. Sometimes referred to as severity on criticality.

Corrective maintenance

An activity that is performed following the failure of an asset, or after the asset has reached a point of unacceptable performance, which is needed to return the asset to service.

Decision support tool

A method or software application that provide a systematic process for decision-making activities.

Deferred maintenance

Maintenance activity which should be carried out but is postponed, usually to save cost, labor and/or material.

Demand management

The active intervention to influence the demand for services and the assets used in the supply of these services to best match available resources to real needs.

Depreciation

The process of spreading the cost of a capital asset over the estimated useful life of the asset. Typically, an expense is charged annually against revenue to write off the cost of an asset due to wear and obsolescence resulting in a reduction in the book or market value of the asset.

ERP (enterprise resource planning)

The process of allowing organizations to manage financial, human resources, and asset information from a central enterprise database and providing integrated management functions to an entire organization. This is typically provided through multiple components of computer software and hardware.

Failure mode

The manner in which an asset can fail to provide the function for which it was installed.

FIS (financial information system)

A software system used to manage the accounting of an organization, typically including general ledger, budget, accounts payable, accounts receivable, purchasing, grants management, payroll activities and fixed asset records.

GASB (Governmental Accounting Standards Board)

An operating arm of the Financial Accounting Foundation that establishes standards of financial accounting and reporting for state and local governmental entities.

GASB Statement 34

A standard released by GASB in June 1999 that establishes accounting and financial reporting requirements for general purpose external financial reporting by state and local governments.

GIS (geographic information system)

A software system used to capture, store, edit, share, display, analyze, and manage data that are spatially (i.e., geographically) referenced.

Infrastructure asset

Long-lived assets that are normally stationary in nature and provide the services of the organization, including pipelines, treatment facilities, and their equipment and appurtenances.

Inventory

An accounting of assets at a facility or site. An inventory includes information such as size and/or capacity, materials of construction, location, installation date, original cost, replacement cost, condition assessment, performance assessment, original service life, etc. Also refers to the actions taken to account for assets at a facility or site. This includes the initial actions to account for assets, as well as ongoing actions to update or improve the asset inventory.

KPI (key performance indicator)

Performance measures that are considered by the entity to be most crucial achieving and maintaining levels of service.

LOS (level of service)

The type and quality of service provided by an entity or an asset of the entity. Levels of service are typically established to consider only the crucial goals of a utility.

Likelihood

The chance of an occurrence; the possibility of something happening. Somewhat synonymous with "probability" but without the support of rigorous statistical analysis.

Life-cycle costs

The total cost of an asset throughout its life including planning, design, acquisition, operations, and maintenance, insurance, rehabilitation and disposal costs.

LIMS (laboratory information management system)

A software system used in a laboratory to track and store the data associated with management of samples, analysis, instruments, standards and other functions such as work flow, labor, materials, etc.

Maintenance

The combination of all technical and associated administrative actions intended to retain an item in, or restore it to, a state in which it can perform its required function.

Master plan

A plan that anticipates by at least 20 years (and may be up to 100 years) future demands on assets within the service area. Typical categories of needs that are addressed in master planning processes include correcting infrastructure deficiencies, meeting growth demands, compliance regulations, customer and other stakeholder expectations, environmental stewardship, etc.

Multi-attribute analysis

A decision support methodology that considers several, sometimes conflicting, factors to quantify the value of an action or project. Also known as multi-attribute utility analysis and multi-criteria decision analysis.

NASSCO

National Association of Sewer Service Companies. A trade association representing companies and agencies involved in wastewater collection system technologies.

NRC-Canada

National Research Council – Canada. The Government of Canada's research and development organization.

Optimization

Finding the best of all possible solutions subject to economic, engineering, political and other constraints.

PCS (plant control system)

A computer-based system that automates the control of mechanical and electrical equipment to manage the operation of a treatment facility or pump station through monitoring, evaluation, and feedback of information.

Performance grade

A representation of an asset's ability to meet its intended purpose.

Performance measure

A metric that reflects the performance of an entity, or an asset of that entity, typically expressed as a ratio of input to output. Performance measures are used to provide information and guidance in achieving level of service targets.

Planned maintenance

Activities that are routinely planned, scheduled, and performed to prevent, minimize, or delay failures and/or shutdowns. Includes preventive maintenance, predictive maintenance, and reliability-centered maintenance.

Predictive maintenance (PdM)

Activities that monitor the operating condition of an asset to determine when maintenance is required. Examples include vibration analysis and performance monitoring. Also referred to as on-condition maintenance.

Present value

A single number that expresses a flow of current and future income (or payments) in terms of an equivalent lump sum received (or paid) today. The calculation of present value depends on the rate of interest.

Preventive maintenance (PM)

Activities that are performed prior to failure of an asset and before the asset reaches a point of unacceptable performance and that will prevent, minimize, or delay failure and/or shutdowns. Examples include adjustments, cleaning, and replacement of minor components.

R&R (rehabilitation and replacement)

Activities that rehabilitate, or replace infrastructure assets.

Reactive maintenance

Activities that are performed on an as-needed basis to address or correct deteriorating performance, failure, or shutdown. Also referred to as corrective maintenance.

Rehabilitation

Restoring an asset's condition to the expected level of service, or "as-new" condition, either by using identical materials and parts or by using some modifications.

Reliability

The ability of an item to perform a required function under stated conditions for a stated period of time.

Reliability-centered maintenance (RCM)

A systematic, disciplined process that involves identification of the causes and effects of asset failures to arrive at an efficient and effective strategy to reduce the likelihood of failure.

Remaining useful life

Estimated life remaining of an asset from current date to anticipated date of retirement, assuming that a reasonable and normal level of preventive maintenance is performed.

Renewal

Rehabilitation and/or replacement.

Repair

Action to restore an item to service after failure or damage.

Replacement

Complete removal and use of another item in place of an asset that has reached the end of its life, so as to provide a similar or agreed level of service.

Risk

A concept and/or value representing the combination of the consequence of an occurrence and the likelihood of the occurrence

Risk reduction/mitigation

Measures taken to reduce risk by either reducing the consequences of failure or the likelihood of failure or both.

SCADA (supervisory control and data acquisition system)

A computer-based system for the monitoring and operational control of facilities through the gathering of data in real time from remote locations.

Service life

Expected life of an asset from installation date to anticipated date of retirement assuming a "normal" level of preventive maintenance is performed. This is the estimated period of time over which a depreciable asset is expected to be able to be used or the benefits represented by the asset are expected to be derived.

SRF (state revolving fund)

Financing programs made available by states pursuant to federal legislation that make low-interest loans available to public water and wastewater systems for planning, designing, and constructing infrastructure.

Stakeholder

An individual or group that has or may have an interest in a project, program, or enterprise.

Top-down approach

An asset management approach that focuses first on a group or set of assets at a high level in an asset hierarchy, such as a system or facility level. It makes the most of existing and available data, as well as institutional knowledge to assess consequence, likelihood and risk, and to identify and select risk mitigation actions.

Unplanned maintenance

See "Reactive maintenance."

Useful life

In relation to a depreciable asset, the estimated total period, from the date of acquisition, over which the service potential of the asset is expected to be used up in the business of the entity.

Valuation

The process of attributing a value to an asset for the purpose of recognizing the asset in financial statements or for a purchase and sale.

Work order

A written instruction detailing work to be carried out.

WRc

A research-based consultancy headquartered in the United Kingdom, focusing on the water, waste and environment sectors.